Microlaunchers

Technology for a New Space Age

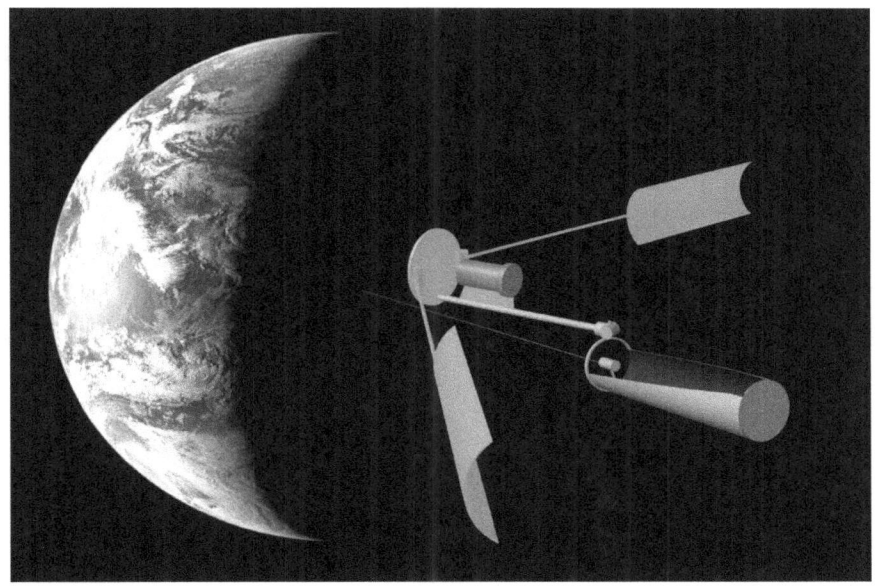

Charles Pooley

Ed LeBouthillier

To those who who have inspired us, we thank you.

TABLE OF CONTENTS

Microlaunchers Introduction

Chapter 1 - Dreams of Spaceflight

My name is Charles Pooley. I was born before the Space Age started and I've watched it develop and change throughout my life. I grew up through the era started by the Soviet Union's launch of Sputnik through the Apollo Era of the 1960's, and through the flights of the Space Shuttle. Living through this era, I was captivated by images of spaceflight, which were seen constantly in the popular media. It motivated me to want to master the technology of rocketry. It has become a lifelong obsession to continue the dreams of spaceflight I seized upon in my youth and extend the reach of humanity beyond Earth.

The idea of extending the sphere of human influence far beyond Earth was a common theme throughout the years of my youth. But somewhere those visions have been lost and we now seem to lack the conviction and direction held in the popular culture back then. We must rekindle that spirit of exploration and expansion not only for the change and excitement it brings, but for the very existence of humanity. We must make space more accessible for all.

Visions of Spaceflight

I first became inspired by visions of spaceflight when I was just a little kid. I remember, in 1950, watching *Tom Corbett, Space Cadet* on a black-and-white television in Dayton Ohio. It was a popular television series continuously showing realistic drama of people exploring far beyond Earth. At about the same time, Flash Gordon buzzed around the galaxy and faced the perils of new and exciting exoplanetary landscapes and adventures.

Destination Moon, a classic of the genre, produced by George Pal in 1950 detailed the perils and excitement of traveling to the moon. I remember watching this and being enthralled by its imagery. In this film, getting to the moon was portrayed as a necessity for American industry and military readiness. This movie foreshadowed the idea of a space race and that idea played prominently in this film.

Forbidden Planet was another cinematic masterpiece that played local theaters in my youth. In staggering brilliant colors, this movie showed a technical future which was both exciting and dangerous.

In 1952, when I was in the fifth grade, Collier's magazine published a series of articles influenced by Werner von Braun. The series, entitled Man Will Conquer Space Soon, was in one of the most widely circulated news magazine and read by many. Visions of reusable winged rocket ships carrying cargo to a spinning wheel shaped space stations floating above the Earth were beautifully illustrated by Chesley Bonestell.

Following the Collier's article, more exciting imagery of space exploration and expansion appeared in Walt Disney's series of movies starting with *Man in Space*, followed by *Man on the Moon*, and finalized by *Mars and Beyond*. This series of films, focusing on how space might be developed, had yet more beautifully detailed and fantastic images. Werner von Braun was the technical advisor on this series of films and appeared in them to explain the technical realities. The imagery was realistic and portrayed as being an eventuality, something that will happen.

It is not my intent to give a broad review of popular TV and movies related to space in reviewing these cinematic stories. I am merely trying to show that the popular culture and imagination in which I grew presumed a fantastic and capable future of space travel and exploration. From these early days I was caught up in ideas of space exploration, particularly in how rockets work, and I wanted to be part of that somehow.

Historical Developments

The historical developments through my youth and onward into adulthood mirrored the pace and excitement of these early films and TV shows. We seemed to be heading towards the fantasy film future I saw through my youth.

Before I was born, Robert H. Goddard had developed the basic technology of liquid rockets. He had published his phenomenal work *A Method of Reaching Extreme Altitudes* in 1910. This established the basic theory on which most of the later work depended. In the years before my birth, he had perfected liquid rockets, their guidance and theoretical underpinnings. Goddard's works are still important and relevant today.

About the time I was born, the V2 was making history both as a weapon of war, and later as a means to reach space. As I grew up, I saw this technology being applied to military weapons and vehicles.

When I was 16 years old, in 1957, I remember explicitly hearing the excited announcement of the Sputnik 1 launch on the evening news. I listened intently and learned that Sputnik would be visible by the naked eye early the next morning. I wanted to see it. Groggy-eyed but brimming with excitement, I woke early, slipped outside into the chilly October pre-dawn and peered Northwards with my binoculars. Sure enough, by the time that my fingers had become chilled from the cold, I looked up and saw a moving light that could only have been Sputnik. The Space Age had started.

As a response to Sputnik, the pace of US space activities picked up. The US government raced to catch up with the public perception of it being behind in the space race. They hurriedly brought out Vanguard which just as quickly blew up in a huge fireball seen by all in newsreel footage. The US rolled out a bigger launcher based on the Juno booster and launched its first satellite, Explorer 1, in January of 1958.

For those of us who followed these developments intently, these were exciting times. All through my high school years, I was inundated by a continuous bombardment of new space development that seemed to be leading, inevitably, towards the kind of space future we all saw in the popular media and knew as inevitable.

There was Gagarin in space, then the US response with Alan Shepard. Then John Kennedy made his famous moon mission speech and John Glenn was launched into orbit.

The pace picked up with Mercury, Gemini, and Apollo. The 1960's were a decade of excited space capabilities development.

These developments were fast paced around me and there were

increasing capabilities before my, and everyone's, eyes. It was common knowledge that we were developing our capabilities for our eventual expansion into space. Additionally, I wanted to be part of that. But, at the time, there was very little chance that I could be involved as anything more than an observer.

Early Involvement

During this era, I was always interested, from the very beginning, about rockets and more about how they worked than the ideas of adventure. The fantasy films and TV, had less interest over me than understanding the technology and capabilities of rockets. I remember studying about them, doing drawings and sketches for engines that I could make. I remember a drawing of a turbopump that I did around 1956.

After I had moved to Los Angeles, California, in 1966, I got involved with one of the two Southern California rocket clubs. The two largest in existence at that time were the Reaction Research Society and the Pacific Rocket Society. I wanted to learn more about the technology of rockets. In 1986 I got involved with the Pacific Rocket Society and a number of us flew what we called "mail flights." We would get postcards, stamped and cancelled in Mojave California and then fly them in our rockets. We would sell them to fund the rocket clubs. Humorously, the postcards that sold best were from rocket flights where the rocket was crashed and crumpled.

In 1993 a project was started to build the first amateur rocket to space. I designed the liquid propellant rocket for that project and we made pretty good headway on building it. We successfully did a static test of the engine. Unfortunately, the project team kind of fizzled apart and progress halted. But, when we tested that engine in June of 1995, it was likely the most powerful amateur rocket engine until about 2008.

Disillusionment

By the time that the Space Shuttle was flying, it was becoming increasingly obvious that something was wrong with the pace and direction of American spaceflight technology development and space exploration. The three-decades long pace and direction was no longer there.

The approach taken with Apollo was so expensive that it was unsustainable. It was carried out with almost blank checks and when it came time to account for the dollars spent, there was little choice but to cut back on what was being done. It had accomplished amazing things, landing men on the moon, but it had essentially emptied out the government bank account to accomplish anything more.

After Apollo was cancelled, there were no Americans going to space for a whole decade. The Apollo program was cancelled in 1971, the last people went to the Moon in December of 1972. It wasn't until 1981 that Americans went to space again. During that time, it was apparent that the vision of space exploration and expansion had withered and died. Dreams of extending human influence beyond Earth orbit seemed completely subsumed by a government program that equated space with the government agency, NASA. The new goal was to go round and round the planet in Low Earth Orbit.

But NASA was also growing more dysfunctional, like a beached whale, incapable of wallowing beyond the shoreline of Earth orbit. Without direction, it immediately started slipping budgets and schedules on the Space Shuttle. We effectively lost the ability to repeat the activity that we had accomplished in the early 1970's of routinely going to the moon.

Space Program Retrospective

The space program from the beginning was a large national policy-driven program of political prestige and the demonstration of power projection. It was expensive, infrequent and centrally planned with a few "boot prints and flags" activities to demonstrate political strength.

It was not about colonizing, exploration or expansion. Directed, incremental growth at smaller scales resulting in increased orbital and extra-orbital infrastructure just wasn't grand enough for attention by the US government. It wasn't sufficient to merely build infrastructure and capability that didn't have an immediate political goal.

I guess, in retrospect, the change occurred during Apollo. Apollo, in reality, being a government propaganda program was designed to show the world the strength and vitality of the grand US military-industrial system. Even as he was setting up the Apollo program for moon exploration, John F. Kennedy is quoted as saying "I'm not that interested in space" unless it showcased American political strength and prowess. Instead of a real exploratory, colonizing space program, we got a bloated, government bureaucracy that was little more than government public relations. Our ability to be involved with space through this kind of program was limited to being government employees and staying in expensive government housing in space.

The space program of the 1950's, 1960 and 1970's was driven purely by military goals and to establish political dominancy against the Soviet Union. Once the Soviet Union conceded the race, the US government had little need for an exploration and colonizing program. In fact, the Outer Space Treaty, signed by the US, Great Britain and the Soviet Union, largely disallowed government ambitions of setting up colonies, effectively killing visions of a whole society projecting itself into space.

I guess we can ask what happened? Just because the government lacked the vision and will to expand into space, didn't mean that we had to stop these visions for ourselves, individually and collectively. As the government programs demonstrated greater capability, we attached our own self-interest in a space future to those activities. We allowed the societal vision for expansion to be subsumed and eventually killed by the government program.

Personal Space

I think it's important to state that our desire to extend humanity beyond Low Earth Orbit is not dependent upon the expectation of governmental involvement and support. Just as easily as we had attached our own interests to the large government programs, we can decide to attach our interests to a non-government form of space exploration and expansion. Just because THEY don't want or need people to go to space for their own needs, doesn't mean that we don't want to go there ourselves for our own needs and wishes. The purpose of our space visions was not so that we could be government employees in space government housing, but so that we could pursue our own space freedom and space future. This is a visionary and revolutionary approach to space expansion.

If we take this visionary approach, we'll have to accept that we can't depend on governmental participation, largely because it is the attractive death call that will just as surely kill our own personal space visions as easily as it has already killed off any state sponsored space expansion. We'll have to avoid chasing the government space carrot for fear of facing the government space stick. Granted, our ability to begin doing space activity will not be as grand as the programs of the 1960's and 1970's initially, but as we build infrastructure and capability, our skills and knowledge will expand.

This may seem grandiose and impossible, but it's not really. Additionally, there is a long history of private space programs extending back through the 1970's. Perhaps the earliest was by German space entrepreneur Lutz Kayser and his Otrag program. Since then, we've seen the development of an increasingly diverse and capable private space effort. Some of the efforts have come and gone. This list includes Starstruck, AMROC, Space Services, Rotary Rocket, Rocketplane Limited, Kistler Aerospace and Beal Aerospace, among others.

But there are others who are already pursuing a separate, private space vision. We are seeing an exciting growth of companies like SpaceX led by Elon Musk, Bigelow Aerospace led by Robert Bigelow, XCOR Aerospace led by its team of founders, Armadillo Aerospace started by John Carmack, Masten Space Systems started by Dave Masten, Garvey Spacecraft and others. These companies (and their leaders) have been showing what a non-governmental space program might look like. So let's consider what a new more expansive private space movement might look like.

Tenets For a Private Space Future

First, a private space program will not be government based. Sure, it may do business with government agencies, but it cannot let government agencies be the sole supporter. It must constantly be doing business with other private companies, including other space related companies. Many people don't realize that doing business with the government is both lucrative and deadly. When the government is one's major customer, then one is led around by government and bureaucratic goals. One has, in essence, accepted the yoke of government control by merely doing business with it. From all manner of its social policies, to its fickle inability towards a constant direction, one accepts this control by doing business with the government. One's bottom line is subject to legislative program approvals and especially subject to massive disruption when the government decides to kill programs.

Without being redundant, a private space program will be private. It will be executed by individuals and the companies that they start for their own reasons. It will consist of individuals, hobbyists, small one-person businesses, family businesses, multi-person businesses, and large private businesses.

A private space effort will be heterogeneous, rather than monolithic. In this way, it will be a movement, rather than a program. It will not be a bunch of people agreeing with each other and pursuing a single large goal (though it might be that at times). Rather, it will be a number of different people pursuing their own interests and goals independently of what those others are pursuing. In this way, it will be somewhat messier because there will not be an overarching direction everyone is pursuing. But in this heterogeneity will come a robustness and expansion far beyond what is possible with monolithic government programs.

There is definitely a necessity to communicate some manner in which this new private space exploration and expansion movement might develop. This book will seek to develop ideas on how small activities can lead to larger activities. It is meant to provide guidance and direction for how a new space era will be brought about. Defining that vision, how it can be realized, what it might look like is the goal of this book.

Chapter 2 - Envisioning a New Future

Considering things being done differently than they have always been done can be unsettling for some. Making them reconsider things can be difficult especially when the new ideas are so very different from common knowledge. If the new ideas contradict the established ideas regarding how things can be done and how they should be done, resistance to the ideas occurs. People often resist change and refuse to consider ideas that are outside of their realm of understanding.

A new private space exploration and expansion movement might seem incredulous to consider for some. That it is private and doesn't need government management might be considered ludicrous. It was once said by William James that "First, you know, a new theory is attacked as absurd; then it is admitted to be true, but obvious and insignificant; finally it is seen to be so important that its adversaries claim that they themselves discovered it." Any radically new idea relating to how space could be developed is likely to also face this progression, especially when it attacks the very foundation of commonly-held ideas if it holds the value of truth.

The task of originating and describing a new movement's ideas can be daunting. Without a clear idea of the kind of vision to start with, those introduced to it may find it hard to relate. For some, thinking about doing things any differently than things are now being done is difficult. How does one organize his thoughts and actions to describing and bringing about a new future?

Approaches to Describing a New Space Movement

There are several ways that one might go about describing a new idea and movement. It is worth considering some of the ways.

One way is to relate it, analogically, to a known historical idea and the movement that followed. Using an analogy, however remote, to what is being proposed provides a bridge for some to grasp and relate similar concepts. We'll call this The Analogical Approach.

Another way is to describe the new movement and its ideas and benefits in a logical and technical detail. When the targeted audience considers them logically, then they might adopt the ideas and bring about the changes. We'll call this The Technical Approach.

Perhaps another way is to just perform the actions that the new ideas suggest and let the results and benefits speak for themselves. We'll call this the Performative Approach.

For the purposes of this book, we shall attempt to do the first two approaches, analogical and logical, with the hope that we and others can perform the third approach. We will also, within our means, do our best to do the Performative Approach. This chapter will consider deriving an analogy for the Microlaunchers idea. The next chapter will derive the Technical Approach to the Microlaunchers idea.

The Microlaunchers Idea: a first introduction

The Microlaunchers idea is that private individuals working singly and together can phenomenally change space exploration and development by starting small and building upon capabilities over time. It also holds that individuals, without government coordination and direction can accomplish far more and amazing things with fewer resources. The key idea is that INDIVIDUALS, not mega corporations and government bureaucracy can and will change the space future for the better.

Following this idea, individuals (and groups of private individuals) can build launch vehicles and spacecraft, engage in space missions, build capabilities, resources and infrastructure and eventually surpass government and bureaucratic approaches. Eventually they will dominate and completely change the course of the future for space exploration and development.

These ideas might seem ambitious. They might seem outlandish to some. Some will scoff and or ridicule them. But those who understand the importance will do the needed work.

The Mainframe Analogy

It is worth considering an analogical approach to how it might be viewed. Let's try an analogy to describing the new approach to space exploration and development. This will be called the "Mainframe Analogy." Here's how it goes.

When computers were first developed, they were large, they were expensive and they were used only by a select few who were largely funded by government efforts. The technology had been developed by government agencies, big corporations and universities to a high level of sophistication. But, like sacred temples they were only staffed by priests and trained acolytes. The average person wasn't welcome to come in and see what they could do in the temple. Laymen had no role in learning about and applying computers during this era.

This analogy pretty closely fits where we are with access to space and the "space program." We may be able to use this model to describe how a new space exploration and development era can develop. But first, let's look at how things changed from the Mainframe Era to the Personal Computer Era.

Some History About the Microcomputer Revolution

Although mainframes still exist, the Mainframe Era has largely been supplanted by the Era of Personal Computers because of the development of microprocessors. This revolution has made computers much more accessible for all of us.

The first microprocessors, which were complete computers on chips, were developed in the late 1960's. Although the military had developed a microprocessor at the same time, one person was largely responsible for introducing the ideas that caused the microcomputer revolution.

Austin O. "Gus" Roche was obsessed with making a personal computer. He originated a design for a microprocessor that would become part of that personal computer. He started a company, called CTC, in 1968, which contracted to another company, known as Intel, to manufacture the microprocessor he called the 1201. Intel was unable to develop it on time, so CTC developed their processor using several chips. However,

Intel eventually released the 1201 under their own name as the 8008 in 1972. Although popular, the 8008 did not spur the revolution that was to come. Intel released the 8080 in 1974; this was the microprocessor that really kicked off the microcomputer revolution.

In July of 1974, a popular technical magazine known as Radio Electronics released an article for the Mark8, and billed it as "your personal computer." In 1975, another technical magazine, Popular Electronics, had an article describing how to build the Altair 8080 computer, based on the Intel 8080 microprocessor. Both of these articles contributed to a wide appreciation of the existence of small, personal computers. Yet another major event was the release of the 6502 microprocessor by MOS technologies also in 1975. The 6800 CPU was another important computer that was released in 1976. The availability of these chips enabled the microcomputer revolution.

There was a flurry of excitement and activity that became a computer hobby movement immediately following these magazines' articles and the releases of these microcomputer chips. Hobbyist groups originated across the country, talking about how to build and program these new personal computers. In March of 1975, for example, the Homebrew Computer Club was started in Menlo Park, California. It was at this club that Steve Jobs and Steve Wozniak first demonstrated the Apple 1.

There was also an explosion of new hobby computer magazines. Creative Computing first published in 1974. Byte Magazine had its first edition come out in September of 1975. Kilobaud, Dr. Dobb's and a host of other magazines were all were started at about the same time. It was a good time to be a computer hobbyist as computer prices became accessible and computer hobby magazines flourished.

From this humble, hobbyist beginning, microcomputers began to become an industry. Obviously, there were the early, small companies like MITS with their Altairs, Southwest Technical Products with their SWTPC 6800 and the IMSAI 8080 by IMS Associates, and other garage entrepreneurs starting companies like Apple. Others started coming into the field like Atari led by Nolan Bushnell. In June of 1977, Apple released the Apple II. Radio Shack soon followed with their TRS-80 Model 1 in August of 1977. IBM, a late-comer to the scene, finally released the IBM PC in September of 1981. But, because they were such a heavyweight contender, they quickly dominated the field.

The Microlaunchers Vision

Using this example of transition from the Mainframe Era to the Personal Computer Era, let's see how we might imagine a change from our current "mainframe era" of space to something much more personal.

Just like the microcomputer chip was the key technology to bring about the change, the Microlaunchers position is that it is launch vehicles, themselves, which is the key technology necessary to change to a more personal space era. Everything else will follow when launch technology is available to a much wider population.

Consider that we are now in the Mainframe Era of space exploration because the "mainframes" of this field are the huge launch vehicles which make space expensive, infrequent and accessible to only a few. In order to bring about the revolutionary change from the Mainframe Space Era to the Personal Space Era, we will need to consider a revolutionary change in how launch vehicles are developed and operated. Doing this will bring about a revolutionary change in access to space and will change how exploration and development are accomplished.

If we use the personal computer revolution as a model for how things can develop in access to space and, consequently, space exploration and development, then we can begin to see a path for how we might change things.

The Microlaunchers Process Model

We saw in the personal computer revolution that key technological changes (i.e. miniaturization of the computer to one chip) brought about widespread interest in computers because they were suddenly available and affordable. Here's a summary of some of what happened.

The first step was *the vision of something new*. A small number of dedicated visionaries saw the possibility for a new and improved way of doing things.

The next step was *the creation or arrival of a new technology*. The first example implementations of this new technology arrived, driven by the visionaries. The new technological approach (miniaturization) allowed the same technology (computers) to be produced cheaper and in larger quantities than before.

Creation created availability. Because of this new technology, which was now available, early adopters saw the possibilities. Old barriers to accessing computers dissolved with the availability of these new microcomputers.

Availability created interest. One or two extremely visible examples quickly created widespread interest. Magazine articles in Radio-Electronics and Popular Electronics made a wide audience aware of a new potential. People saw new possibilities with these new, small computers and they realized that they could get involved.

Availability and Interest created Community. From the nascent interest and availability, a hobbyist community quickly developed. This community quickly started tinkering and experimenting with this new technology. It was sometimes known as "homebrewing" as you saw earlier. This interest resulted in the quick development of a number of magazines and clubs. New tools and techniques quickly spread among these hobbyists. New technologies (in the form of new chips) were written about in magazines, this created more interest and activity which then resulted in more articles and magazines. Hobbyist designs of computers and support equipment flourished during this time but at a hobbyist level.

Availability, Interest and Community created Markets. The next stage of the revolution was the establishment of increased small businesses to support these early hobbyists. People were buying components, assemblies and whole systems.

Availability, Interests, Community and Business Support made a sustainable revolution. All of these things interacted to create a sustained and self-supporting revolution.

So, let's apply this process model to changing access to space.

1. Vision - those of us who desire and envision a new, better future where space is routine, cheap and accessible are setting about on a new approach to space launch vehicles: microlaunchers. We envision starting with the smallest reasonable launch vehicles to fulfill hobby and personal needs and grow our capabilities as availability becomes wider.

2. New Technology - the new technology in the Microlaunchers approach is the miniaturization of launch vehicle technology. This will allow it to become affordable and available to a wider audience.

3. Availability - with the availability and affordability, early adopters can apply the technology of miniature launch vehicles to solve their own problems.

4. Interest - as early adopters show that new things can be accomplished for less money, there will be greater interest in using microlaunchers for yet more projects.

5. Community From Availability and Interest - as more people see the availability of small launch vehicles and become interested in them, a community will develop. This will have its own media for communicating new techniques, applications as well as improvements.

6. Markets arise from the community - the early community will need components and assemblies and entrepreneurs will strive to serve their needs. As this grows, there will be more interest, capability, and availability.

7. A sustainable revolution exists - Access to space will be forever changed by the new community. The microlauncher market will sustain itself and lower access to space.

The following timeline shows the steps in the microcomputer revolution and the suggested microlauncher revolution.

The Microcomputer Revolution   The Microlaunchers Revolution

VISION: VISION:
   Personal Computers   Personal Launch Vehicles

NEW TECHNOLOGY: NEW TECHNOLOGY:
   Single Chip Microprocessor   Small Launch Vehicles

AVAILABILITY: AVAILABILITY:
   First Hobby Computers   First Hobbyist Microlaunchers

INTEREST GROWS: INTEREST GROWS:
Hobbyists and Magazines Flourish Hobbyists and Magazines Flourish

COMMUNITY DEVELOPS: COMMUNITY DEVELOPS:
   Growing Mutual Support Groups   Growing Mutual Support Groups

MARKETS ARISE: MARKETS ARISE:
   Commercial Computers   Commercial Vehicles and Parts
   Widely Available   Widely Available Off the Shelf

SUSTAINABILITY: SUSTAINABILITY:
   Large Corporations Dominate   Large Corporations Make
                                     Vehicles and Parts Off the Shelf

Chapter 3 - Microlaunchers Technology Introduction

In our introduction to the Microlauncher Process Model, it was identified that one key technology (or collection of technologies) was necessary for the envisioned change. This technology was miniaturized launch vehicles, or microlaunchers. But, are microlaunchers even possible? What are the constraints that must be addressed and overcome? We will begin reviewing some of these ideas here.

When told that launch vehicles can be quite small, even people educated in the field of rocketry often scoff at the idea as impossible or impractical. They will ask for evidence or even cite examples of why it is not possible.

Just because no one had tried to create a very small launch vehicle does not automatically mean that it is impossible. It merely means that no one has tried or has been successful. There *is* a practical lower limit on how small a launch vehicle can be, but that limit sets the size of the rocket far smaller than most people realize. It is only because of their lack of familiarity with the scale of the forces affecting the launch vehicle and the means to overcome those problems that even those educated in rocketry consider them impossible to overcome.

Historically, the smallest successful launch vehicles to date have weighed about 10000 kg at takeoff. For the revolution that is advocated by the microlaunchers vision, we are talking about vehicles that are one to two orders of magnitude smaller than this. Therefore, we are talking about vehicles in the range of 100 kg to 1000 kg in takeoff weight. If these can be made affordably, then the technology to cause the microlaunchers revolution will exist.

Historical Examples of Small Launch Vehicles

It is worth examining the historical examples of small launch vehicles to establish an incontrovertible lower bound on successful launch vehicle size.

Lambda 4S

The smallest launch vehicle which successfully placed a satellite into orbit to date is the Japanese Lambda 4S launch vehicle. In 1970, it successfully placed the satellite Osumi into orbit. The Lambda 4S was a four stage solid propellant vehicle weighing about 9399 kg at liftoff. Its total length was about 16.5 meters and its largest body diameter was about 73.5 cm.

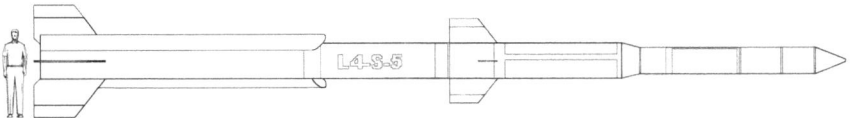

Vanguard

The US Vanguard launcher is another small launcher worth looking at. It was the second-smallest launch vehicle which successfully orbited a satellite. The Vanguard was a three stage launcher with the two lower stages using liquid propellants while the third stage used solid propellants. It weighed about 10341 kg at takeoff. It was about 23 m long by 1.14 m in diameter.

Getting into Orbit

It was Sir Isaac Newton who was one of the first people to mathematically consider what it took to get into orbit. In one thought experiment, he imagined a cannon on a tall mountain above the atmosphere shooting a cannonball horizontally to the Earth's surface.

He imagined that if the cannon shot a cannonball at a low velocity out of the gun barrel, the cannonball would fly out so far and hit the Earth downrange (point A in the figure below). Considering that if the cannonball were fired out a bit faster, he knew that it would go even farther downrange (point B in the figure below).

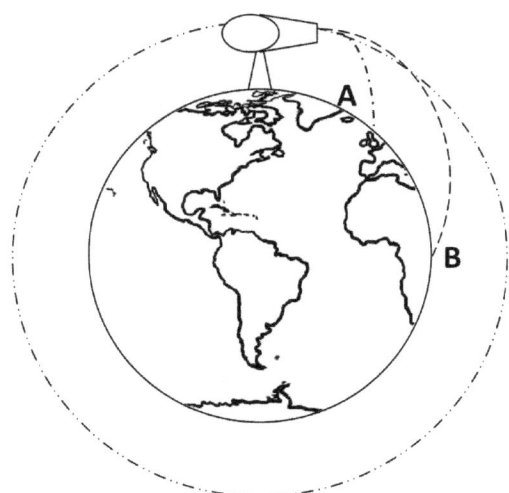

But if the cannon fired the cannonball at a particular velocity, it would go all around the way around the Earth and come back and hit the cannon. He also considered that if the velocity of the cannonball were fast enough, it would escape from Earth.

This is the basic idea of being in orbit. At a certain velocity, the forward motion equals the downward motion due to gravity and the object will continue around repeatedly in an orbit.

A launch vehicle is a rocket that raises the payload to a certain altitude and velocity necessary for the kind of orbit desired. A launch vehicle is able to put a satellite into orbit around the Earth, the Sun, or another planetary body.

The Physics of Launch Vehicles

A launch vehicle, like any vehicle that flies through the atmosphere, is subject to four fundamental forces: lift, drag, thrust and gravity. The following diagram shows each of these forces in relation to each other and the launch vehicle.

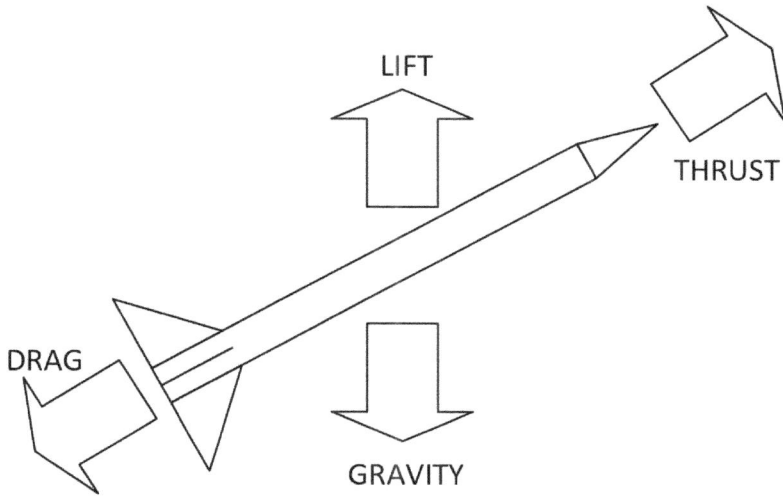

Thrust is what moves the rocket forward. It is produced, usually, by a rocket engine.

Drag is the force produced by air resistance. It works oppositely to the direction of flight and slows down the vehicle. Generally, the faster a vehicle travels, the larger the drag force which is trying to slow it down.

Lift is the force produced by wings and similar aerodynamic surfaces. In an airplane, it is produced by the wings. With rockets, since they usually have no wings, lift is generally negligible. In fact, lift is generally something to be avoided since it also comes with increased drag.

Finally, gravity produces a force which pulls the launch vehicle back towards the Earth.

A rocket uses a rocket engine to produce thrust to overcome the influence of gravity and drag.

In order to put a payload into orbit, a launch vehicle must be able to accelerate itself and its payload up to orbital speed. Orbital speed is not an insignificant velocity, being around 7.8 km/s. Since gravity is constantly pulling the launch vehicle towards the center of the Earth's mass, taking too long to attain orbit also can cause too much fuel to be used. Most launch vehicles seek to accelerate their payloads within a period of 2 minutes to 6 minutes. Therefore, most launch vehicles will accelerate from standstill to orbital velocity at a rate of from 2 g's to upwards of 20 g's (where 1 g is the acceleration felt at the Earth's surface, or 9.8 m/s^2).

How Small Can a Microlauncher Be?

Knowing that examples of the smallest successful launch vehicles to date weighed about 10000 kg at liftoff, what issues might a smaller launch vehicle face? We need to go back to the forces acting on a rocket to know what factors affect the lower bound on launch vehicle size.

For a launch vehicle that must fly through the atmosphere to space, it must be able to accelerate its payload to about 7.8 km/s. In order to do that, it must generate sufficient thrust and overcome aerodynamic drag and the force of gravity. Of the four forces that it faces, only aerodynamic forces significantly change depending on the size of the launch vehicle. The acceleration of gravity is the same, so the force of gravity scales with the launch vehicle mass. The thrust of the vehicle is dependent upon the rocket engine that it has and can be scaled coincident with the size of the vehicle.

The force due to aerodynamic drag, as we change the vehicle scale, is primarily dependent upon one factor: the ratio between its mass and its frontal area (and the coefficient of drag, which is a factor defining relative vehicle drag). Since a smaller vehicle has a relatively larger frontal area for its mass, it will experience relatively more force than a larger vehicle. Therefore, the primary factor that is different in a smaller launcher is a relatively larger aerodynamic force. Can the nature of this force be so great as to disallow a small launch vehicle in the microlaunchers range from being possible? The answer is no. Actually, the answer is "it depends" and it depends on how small the mass of the vehicle is and its frontal area. There is a lower limit at which aerodynamic forces become unwieldy but that lower limit is still a pretty small vehicle. But we'll address all of these factors in greater detail later.

A Deeper Look at Drag Effects

Understanding the relationship of air drag is vital to understanding the lower limit on the size of a launch vehicle. Rather than introduce this complex idea of aerodynamic scaling with rocket shapes, let's simplify the conception of this force by using a simple shape like a sphere shot out of gun. In this experiment, there are a wide range of sphere sizes, all shot out of guns at the same velocity. All of the spheres are made from the same material, and only their diameter changes. Once they leave the gun barrel, then they will encounter aerodynamic drag.

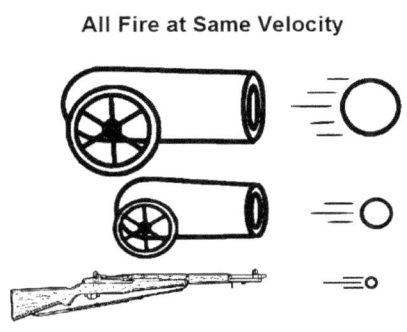

All Fire at Same Velocity

1X Reference Object
Diameter = 1 meter
Area = 0.787 sq meters (1X Reference Area)
Mass = 1X Reference Mass
Deceleration = F/M = 1/1 = 1X

0.1X Scaled Object
Diameter = 0.1 meter
Area = 0.01 of Reference Area
Mass = 0.001 of Reference Mass
Deceleration = F/M = 0.01/0.001 = 10X

0.01X Scaled Object
Diameter = 0.01 meter
Area = 0.0001 of Reference Area
Mass = 0.000001 of Reference Mass
Deceleration = F/M = 0.0001/0.000001 = 100X

With this simplification, we can identify the effect of aerodynamic drag on these spheres ranging from 1/2 meter diameter down to very small (like a BB size). If we could measure the speed of each of these spheres at two points as they leave the barrel, we could identify the rate of deceleration due to aerodynamic drag. This provides a good analog to understanding the scale of aerodynamic forces on small launch vehicles since the same factors of mass and frontal area are in play.

The following graph shows the deceleration experienced by these spherical objects traveling through the air versus each sphere's diameter. This deceleration is the product of the forces created by air drag, F_{drag}, and its effect on the mass, m, of the projectile. Due to these two influences, the overall deceleration is very different for different sized objects.

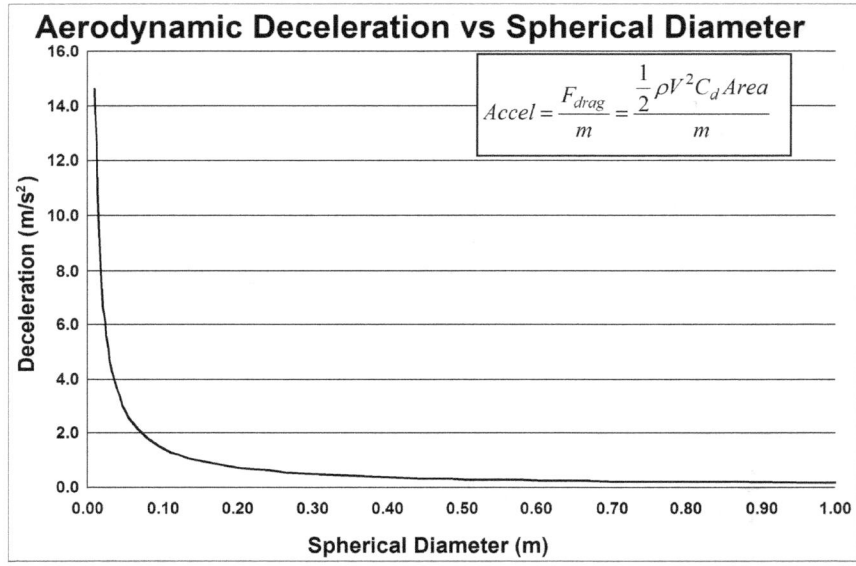

As can be seen, as the sphere diameter is decreased, the deceleration due to drag is increased in a very non-linear way. If one decreases the size of the sphere by ½ (say from 1 meter to 0.5 meter), there is very little noticeable change in the deceleration. But as one decreases from 0.5 meter down towards 0.25 meters diameter, there is more noticeable deceleration. As one goes from 0.25 meter diameter down towards 0.10 meter diameter the deceleration is far more pronounced.

If the drag on the 1 meter sphere is 1X (Drag1), then the drag for the 0.1 meter sphere is 0.01X (Drag2) and for the 0.01 meter sphere, it is 0.0001X (Drag3). But, if the force of drag is divided by the mass, since the mass goes down significantly more, Drag1/Mass1 = 1X, Drag2/Mass2 = 10X and Drag3/Mass3 = 100X.

Now, spherical bullets aren't rockets, but the dimension of the forces is the same with rockets. It's just easier to calculate the area and weight of a spherical object for the purposes of illustrating scaling effects of drag on rockets.

Put in the orders of magnitude change we've talked about with microlaunchers (from 1 to 2 orders of magnitude smaller than the smallest successful launchers to date), we can see that there will be a significantly greater deceleration due to aerodynamic effects. For example, going down one order of magnitude on the graph above, from 1 meter diameter down to 0.5 meter diameter, we see that drag deceleration is increased about 10 times. Going down two orders of magnitude from 1 meter diameter down to 0.01 meter diameter, the drag deceleration is about 100 times greater. The effect appears to be a 1/x relationship between size and aerodynamic drag deceleration. This means that the proportion of propellant required to overcome drag is a greater amount for smaller vehicles. This will limit how small a stage can be.

Therefore, one of the key aspects of making microlaunchers viable is developing ways to deal with the significantly greater effects of air drag. One of the most obvious ways is to just make the frontal area relatively smaller than a strictly linear scaling. This means making the rocket's body diameter smaller and consequently increasing the length of the launcher relative to its diameter. We will look at this issue in greater detail in a later chapter.

Another approach to dealing with air drag issues is to trade off gravity losses for aerodynamic losses. This is done by tailoring the acceleration during ascent in the atmosphere. By starting off a bit slower in the lower atmosphere, some fuel will be lost to fighting gravity, but significantly less aerodynamic deceleration would be experienced. By properly controlling this, we can reduce the overall fuel wasted fighting the atmosphere.

Yet another approach to minimize drag is to make the first stage's ascent trajectory be vertical. This will cause it to minimize the flight through the atmosphere. Once the vehicle clears the atmosphere (or reaches a point where the atmosphere is thin enough), then the upper stages can be used to impart the needed lateral velocity.

Scaling Down Launch Vehicles

So, if our goal is to make launch vehicles in the range of 100 kg to 1000 kg at takeoff, what will it take to do that? Obviously, all of the parts of the launch vehicle will have to be made smaller while maintaining their performances at sufficient levels. It is also hoped that in scaling down the launch vehicle, that launch costs will also scale down (though maybe not by the same factor but hopefully by some significant factor).

Let's look at an example. Suppose we use the following weight budget for two different rockets, one at the lower scale of microlaunchers (100 kg) and one at the upper scale (1000 kg). These budgets are derived from an actual sounding rocket, the Aerobee 150A, so they represent realistic budget allocations for vehicles of this class.

Subunit	Percent	100 kg Total Mass	1000 kg Total Mass
Payload	12.9	12.9 kg	129.2 kg
Nose Cone	1.1	1.1 kg	11.0 kg
Rocket Engine	2.1	2.1 kg	20.7 kg
Tanks	10.5	10.5 kg	105.3 kg
Oxidizer	49.0	49.0 kg	490 kg
Fuel	19.1	19.1 kg	191.4 kg
Pressurant	0.5	0.5 kg	4.5 kg
Propellant Feed	0.3	0.3 kg	2.6 kg
GNC	1.3	1.3 kg	12.5 kg
Structure	3.3	3.3 kg	32.7 kg

Examining the GNC (Guidance, Navigation and Control system), we begin to see some important trends. Even though the GNC performs exactly the same function in the 100 kg launcher versus the 1000 kg launcher, there is much more absolute mass allocated for it in the larger rocket. In fact, it might be difficult to make the GNC meet the mass budget of the smaller rocket. Some off-the-shelf parts (like electronics) aren't available in physically smaller sizes. Manufacturers offer only a limited range of sizes and that availability sets a lower bound on off-the-shelf components.

This mass allocation budget illustrates at least one problematic aspect of scaled-down rockets: scaling introduces unique problems on some of the components. It might be necessary to change the mass budgets for smaller vehicles to deal with certain difficulties when selecting small, lightweight components.

Small Rocket Performance Examples

We have established the scale of aerodynamic effects for small vehicles. But, rather than think that it makes it impossible for them to have the necessary performance, let's look at historical examples of effective small launchers. We will look at two small sounding rockets that have demonstrated sufficient performance to work as microlaunchers: the Super Arcas and the Astrobee D. These examples prove that the aerodynamic forces are not unconquerable.

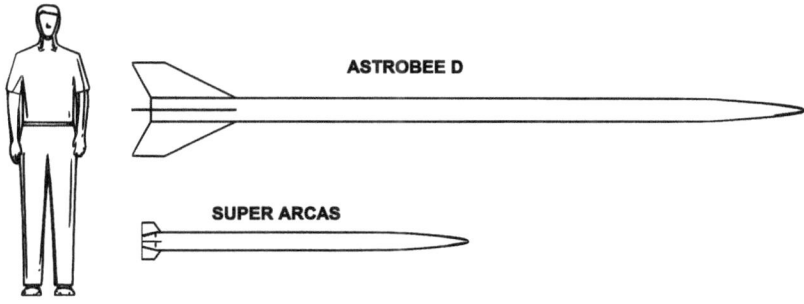

ASTROBEE D

SUPER ARCAS

The Super Arcas was a small sounding rocket able to reach altitudes as high as 100 km with a payload of 3.6 kg. It had a diameter of about 11.43 cm and a typical length of about 2.7 meters. Four fins stabilized it in flight. It utilized an end-burning solid propellant motor with 25 kg of propellant which produced an average thrust of 1445 Newtons for a total burn time of 40.2 seconds. It weighed 37.5 kg without the payload and 12.5 kg empty.

Based on the above information, we can establish the aerodynamic drag that it faced. From simulations and reported figures, it appears that the theoretical maximum velocity was 2175 m/s. Gravity slowed the vehicle down by about 394 m/s during the motor burn. At the end of the thrust duration, the vehicle was traveling about 1307 m/s. Therefore, aerodynamic drag slowed the vehicle down by about 474 m/s. This is about 10 times what the Saturn V experienced during its flight. This relatively low aerodynamic speed loss was because of the relatively high length to diameter ratio of 23.5:1 and because the mass didn't scale down at the same ratio.

The Astrobee D was another sounding rocket slightly larger than the Super Arcas. It had a diameter of 15.24 cm and a body length of about 397.5 cm. It was able to lift a payload of 15 kg to an altitude of 100 km. It utilized an end-burning solid propellant motor with 60 kg of propellant which burned for about 20 seconds. It weighed about 93.4 kg without a payload and weighed about 33.1 kg empty. It had a 2 second boost thrust of 17 kN and an 18 second sustain thrust of 8 kN.

Based on published information, the theoretical maximum velocity would be about 1922 m/s. Gravity would slow it down by about 183 m/s in its thrust duration. At the end of the thrust duration, it would fly about 1292 m/s, so the aerodynamic loss is calculated to be about 447 m/s.

Again, this example shows that small rockets can reach the edge of space with substantial payloads despite the significantly larger aerodynamic drag they encounter. The length to diameter ratio of the Super Arcas was about 23.5:1 while that of the Astrobee D was about 26:1.

So, small size is not necessarily an indicator of an inability to reach space or carry appreciable payloads to the edge of the atmosphere. This is the kind of performance that microlaunchers require. As was seen, by making the length to diameter ratio higher than 20:1, these vehicles overcame aerodynamic drag sufficiently enough to reach space.

As a general rule, there is a limit to how long you can make a rocket relative to its diameter. Beyond some points, the vehicle is too weak if it is too long and skinny. In other words, it bends too much and breaks. Therefore, in general, microlaunchers should seek to maintain a length to diameter ratio close to 20:1 although they may go as high as 30:1 length to diameter ratio. Higher ratios will require more careful analysis of structural issues. But, these high lengths to diameter ratios will result in lower drag needed to meet performance requirements.

Microlaunchers' Performance Requirements

Microlaunchers, in order to meet their goals of putting small payloads into orbit or escaping Earth, require higher levels of performance than the level amateurs usually build their rockets at. A major strategy of perfecting microlauncher technology is to make smaller launch vehicles increasingly more efficient so that they approach the efficiency of larger commercial launch vehicles. This usually means adapting the propulsion technologies used in commercial launch vehicles and applying them to these smaller scale rockets.

There are two major metrics to measure the efficiency of rockets: Specific Impulse and mass ratio. Together, these two metrics define the size of a rocket able to go a certain speed. Therefore, in order to get smaller rockets, these two metrics must be comparable to the general state of the art for launch vehicles in general.

Specific Impulse (often abreviated I_{sp}) is the rocket equivalent of Miles Per Gallon (MPG). Whereas, in a car, MPG measures the distance a given volume of fuel produces for a ground vehicle, for rockets, *Specific Impulse* measures the time duration that one mass unit of rocket fuel can produce one unit of thrust. The more efficient a rocket engine is, the longer duration it can produce one unit of thrust for one mass unit of propellant. The most efficient engines can produce one kilogram-force of thrust (9.8 Newtons) for about 450 seconds using one kilogram of propellant. Black powder rocket motors, like those used in hobby rockets generally produce one kilogram-force of thrust (9.8 Newtons) for 80 seconds using one kilogram of propellant.

The *mass ratio* is a metric for determining how light (or heavy) a vehicle is. It compares one subsystem's mass against another subsystem's mass. There are many different mass ratios used, but a very useful one, the propellant ratio, relates the percentage of propellant mass to the mass of the propellant plus the empty vehicle. In this case, the propellant mass (Mp) is divided by the vehicle empty mass (Ms) plus the propellant mass (Mp) to get one type of ratio. Alternatively, there is another popular mass ratio figure, Mf/Me, which represents the full loaded value of the vehicle, Mf (which includes the empty stage, the propellant and the payload), divided by the vehicle without propellant, Me (which includes the empty stage and the payload). Both of these are important ways of specifying the mass performance of a rocket stage. We will discuss them again in a future chapter.

Because of the need for higher specific impulse and lower mass ratios, microlaunchers are envisioned as being multistage liquid rockets with medium performance rocket engines and fairly lightweight structures. The following table illustrates recommended performance figures for microlaunchers.

Parameter	Stage 1	Stage 2	Stage 3
Sea Level I_{sp}	200s - 230s		
Vacuum I_{sp}	250 s	306 s	306 s
Mass Ratio (Mf/Me)	2.5	4.6	4.6
Propellant Ratio (Mp/(Ms+Mp))	80%	85%	87%

These are ambitious but realistic goals, even for dedicated amateurs. Given these kinds of performances, microlaunchers vehicles will be quite small but have very high mission performance. To give an idea of the scale of the vehicles being talked about, the following figure shows two advanced designs for perspective.

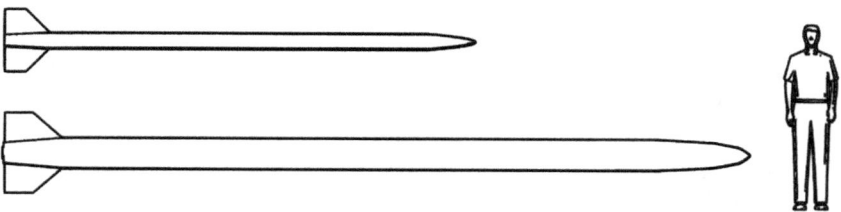

Another unique aspect of microlaunchers is that the first stage is seen as taking a different role than for most commercial launchers. Most commercial launchers' first stages have two roles: raise the altitude of the upper stages above the atmosphere and impart some significant lateral velocity to them. However, because of the serious impact of aerodynamic drag on small launch vehicles, it is envisioned that the first stage only has one role: raise the upper stages above the atmosphere without significant lateral velocity. This is seen as a strategy to deal with the problems of aerodynamics.

Microlaunchers Are About Escape Missions

Finally, one aspect of microlaunchers that makes this approach very different from a lot of other approaches is an emphasis on escape missions. This means deep space missions with very small payloads. Because these are vehicles with greater than Low Earth Orbit capability, they can certainly do orbital missions. But, by focusing on missions beyond Earth orbit, it is envisioned that this opens up the capability for real science and explorations missions even for amateurs. It is also envisioned as a facilitator for getting launch licenses because missions beyond orbit have no collision hazards.. Because of this unique approach, we will also discuss ideas for missions and technology supportive of deep space exploration.

Chapter 4 - How to Bring About the Microlaunchers Future

We've taken a first look at the kind of future that microlaunchers can bring as well as the basic ideas behind the launch vehicles that make it happen. It is worth considering in greater detail how the microlaunchers revolution will take place.

Microlaunchers - A Non-Analogical Perspective

In earlier chapters, we talked about microlaunchers and the changes they could bring by using the microprocessor revolution analogy. Analogies are useful for providing a big picture perspective on what they describe, but they often break down in describing details. To address some of the details, we can consider what the microlaunchers revolution suggests without using analogy. As a process the microlaunchers revolution would consist of the following developmental stages:

1. Go for Escape Missions rather than Orbital
2. Start Producing Launchers and Payloads at the Smallest, most affordable level possible
3. Develop a Hobbyist and DIY base
4. Create A Startup Venture Environment With Advanced Technologies
5. Extend Development to a Sustainable Professional Support Base
6. Extend and Develop Capabilities leading eventually to human-scale launchers

We shall examine each of these steps in greater detail in the next few sections.

Going For Escape

Although there are a lot of people with an interest to get a satellite into Earth orbit, a key idea of microlaunchers is to go beyond Low Earth Orbit (LEO). It's not that orbiting satellites aren't interesting; it's just that there are a lot of impediments for hobbyists and amateurs to be able to do orbital missions.

First, LEO is not space exploration any more. Just about anyone can download images of Earth from orbiting satellites on the World Wide Web nowadays. Exploration begins beyond Earth's sphere of influence. The moon, asteroids, the Sun, the Stars and other planets are within the grasp of microlauncher developers if they look beyond LEO.

Second, LEO is becoming increasingly regulated by the governments of the world. There is a lot of expensive equipment in orbit that precludes the ability to experiment or innovate at LEO. This is because there is too much of an opportunity for collision hazards that can damage more than the experimenter's hardware. Getting a spacecraft into a precise orbit requires extremely delicate control capabilities; early microlaunchers may not be able to prove their ability to get specific orbital trajectories. Therefore, trajectories that take experimental vehicles away from costly orbital assets will be preferred by regulators. It will be easier to get a launch license if one can prove that they don't threaten any existing orbital assets.

Another reason against simply going for LEO satellites is that something in low orbit is out of sight most of the time. It takes at least ninety minutes for a satellite in LEO to return above the ground station, more if the orbit is precessing. Sometimes, LEO orbits can take months before they return above a particular ground point.

On the other hand, something on an escape trajectory or in orbit around a planetary body or asteroid will be in a receptive position for hours per day. Using radio signals, one can receive data during the day and night but this requires large, expensive ground stations. Using laser communications, the data can be received at high rates with relatively little equipment. Laser communications is the preferred method for microlaunchers communications because it is currently an unregulated electromagnetic band. We will discuss these ideas more in a later chapter.

Starting Small

One of the great benefits of the microlaunchers revolution is the idea that physically small launchers and payloads can make a big difference to the exploration and development of space. Because of the small physical size, some of the costs of development are drastically reduced.

Because the vehicles are, dimensionally, not any larger than existing High Power hobby rockets, much can be adapted from other hobby developments, although the key propulsion and propellant feed technologies will have to be developed without being able to be borrowed from other fields. Let's quickly review some available and useful techniques and technologies that are ready to be used by microlaunchers.

RC Airplane and Car hobbies have, over the years, developed very effective and affordable technology that can be applied to the development of microlaunchers. Small servos, receivers, and highly efficient motors and motor controllers are now very affordable and available. These are a tremendous resource to advance microlauncher development.

Because of the preponderance of small consumer electronics devices, there are many amazing, small electronic components that can be used in microlaunchers and their payloads. Small motors are used in all manner of consumer devices, from cell phones to hand drills. There will be need of these small motors to control numerous devices and mechanisms in the launchers and their payloads.

Small CCD and CMOS imagers are also widely available since they are used in many consumer electronics devices. These can be used in guidance systems, event recorders, and in imaging systems.

There are a number of useful integrated circuit devices that microlaunchers can utilize. From small microprocessor chips, to power control chips, memory chips and all other parts needed for guidance, communications, telemetry and data storage mechanisms.

An additional important resource is the availability of small, cheap powerful diode lasers and very sensitive avalanche photodiodes at reasonable prices. These two components can produce highly sensitive long-range communications systems between a ground station and a spacecraft.

Since it is highly unlikely that small pump-fed liquid propellant rocket engines will be available immediately, there are simpler techniques which are sufficiently capable and recommended. Specifically, the use of pressure-fed propellant feed systems allows sufficiently efficient microlaunchers.

Rocket engines require that their propellants be fed into the combustion chambers at pressures higher than that at which they are being burned. Most large, commercial liquid rockets use pumps to bring the propellants to high enough pressure to be injected into the combustion chamber. But pump systems take more development time, so early microlauncher developers can utilize a pressure fed system to feed the propellants. The following diagram represents two of the recommended pressurization schemes.

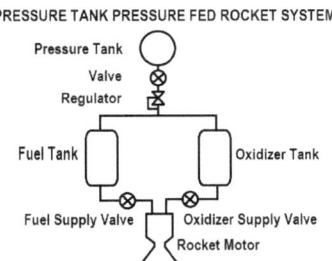

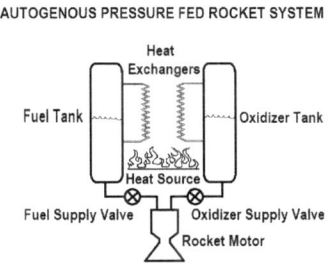

In the left diagram, a separate pressurant tank, like a paintball pressure tank, is used to pressurize the propellants and feed them into the rocket engine combustion chamber. This is known as a Pressure Tank Pressure System. It has the advantage that the pressure can be carefully and easily controlled with a regulator.

In the right diagram, either one or both of the propellants flows through heat exchangers and is vaporized, producing the necessary pressurized gas to feed the propellants into the rocket engine combustion chamber. An obvious heat sources is the rocket engine itself. This system requires somewhat more advanced techniques but can be lighter.

Although not shown in the diagrams above, it is also possible to use a solid propellant to generate the pressure to feed the propellants. This too can be small, simple, light and efficient.

All of the above mentioned technologies and techniques will allow small, lightweight and sophisticated mechanisms in microlaunchers comparable to large, commercial launch vehicles.

As microlaunchers are built and developed, they will initially have low mass ratios because the skills and techniques of building lightweight subsystems will not automatically exist. However, key to making microlaunchers small is attaining good mass ratios. By developing unique techniques and borrowing some from other fields, microlauncher developers will be able to create better mass ratios and smaller, more capable vehicles.

Because of their small size, it will be possible and desirable to launch often and repeatedly. Although reuse of the stages is not being advocated (or advocated against), it may be possible. But, in any case, because of the relatively low cost and size of these launchers, it should be able to launch them as easily as High Power hobby rockets are today, at least to test and perfect them.

Developing a Hobbyist and DIY Base

In the early stages of the development of microlaunchers and their constituent technologies, the most important components will not be available widely or commercially manufactured. It will be up to experimenters, hobbyists and Do It Yourselfers (DIYer's) to develop their own microlauncher subcomponents. But, by maintaining contact with other microlaunchers developers, ideas and techniques will be shared and development sped up.

In order to increase the rate of development of key components, microlauncher developers should seek to build interest and show off their accomplishments with each other. As we saw with the microcomputer revolution, small hobbyists groups coming together to share ideas and accomplishments played an important role. Along with these hobby groups, a number of newsletters and magazines began stimulating interest and involvement. The same needs to happen with microlaunchers.

There are many ways that the interest can be built and sustained. Competitions are one way that people can be motivated to accomplish key goals such as improving mass ratios for a given performance, or building a rocket engine of a certain performance specification.

Groups of microlauncher enthusiasts can hold conferences where ideas and equipment can be displayed, traded, sold, and bartered for. These can be their own microlaunchers conferences or tracks at other conferences.

A slew of microlaunchers hobbyist and trade magazines needs to develop. Imagine receiving a monthly microlaunchers magazine with the latest developments from your favorite innovators. That would surely spark interest to accomplish more in a lot of people. There are a slew of other ways to get others interested and excited in microlaunchers development, including magazines, books, newsletter, web pages and social media pages. There can even be online design repositories and open source collaboration projects to develop key components and ideas.

This continuous collaboration at the hobbyist level will build a growing and dedicated hobby base that will grow to the next level.

Startup Ventures

As the hobbyist base develops, there will be increasing opportunities for microlauncher experimenters to sell their hardware to other microlauncher enthusiasts and companies. Again, following the microcomputer revolution, this would be an early stage where there is growing demand for specific necessary parts such as rocket engines, feed systems, plumbing components, control components, guidance systems, payload systems etc.

Capable microlaunchers enthusiasts will see opportunities to produce and sell parts that they've developed experience in. Perhaps there will be a few companies selling whole microlauncher systems at well-funded hobbyist levels. This might seem ridiculous on initial consideration, but who would have thought that there would be, today, a market for hobbyist jet engines? In the same way, advanced rocket engines have research, aerospace, science and military applications once their basic capabilities are established and their performance and costs controlled.

Universities will see immediate opportunities for their students to be involved in actual exploration projects using microlaunchers and the small spacecraft launched by them.

Developing a Sustainable Professional Support Base

There will be a time that microlaunchers can be a significant business market. Using the microcomputers analogy, there was a time that companies realized there was significant money to be made in microcomputers and companies like Commodore, Atari, Apple and Microsoft arose. In the same way, there may be significant companies making revenues from building microlaunchers, microspacecraft and various components for them. Of course, the microlaunchers market can not be as big as the microcomputer market, but it can still be a significant niche market equivalent to at least hobby jets. The actual microlaunchers market size is likely to be several times larger than hobby jets and helicopters, for example. This is because, if only a small percentage of universities were interested in these kinds of vehicles, this would be still be several thousands per year.

Extending to Human Scale Launchers

There may come a time where microlaunchers and microspacecraft can eventually lead to a hobbyist scale human launch capability. By this time, microlaunchers will have created the opportunity for widespread launch components and systems based on cheap, mass produced rocket subsystems.

The easier commercial availability of these subsystems will allow off-the-shelf parts to be combined into more and more capable launchers. For example, it might be possible to cluster many smaller launchers into a single larger launcher suitable for launching even human payloads.

In any case, the wider technical capabilities of the microlauncher support industry could quickly adapt its knowledge, skills and infrastructure to larger projects which might include human missions.

The Beginnings of the Microlaunchers Revolution

Now that we've covered the overall plan of how the microlaunchers revolution should develop, we might think about the beginning parts that can make it happen. There are obvious key technologies required. The electronic parts needed for computers, guidance, control, imaging, communications already exist. They are part of the microelectronics revolution that we've seen in the past 50 years or more. Cubesats have become widely popularized in universities and industry. They're already showing an established industrial base.

The key parts needed for microlaunchers are small efficient rocket engines and propellant feed systems becoming widely available. The first game-changing technology will be modular rocket engines, feed systems and plumbing components at sufficiently high enough performance to satisfy microlauncher mission requirements. The first example of the impending revolution would be when you see an advertisement for a suitable rocket engine available off-the-shelf at a price that hobbyists can afford. As stated earlier, these will be pressure fed rockets at first because that's the simplest, requiring the least development. But complete pump fed rocket engines in the microlaunchers mass ranges can be developed. Pump fed rocket engines are no more complicated than the hobbyist jets that are flying today. The technology is suitable for mass production at prices similar to hobby airplane motors and jets.

There will likely be three, maybe four, classes of engines needed. The following table shows the likely classes of motors and their application.

Parameter	Stage 1	Stage 2	Stage 3	Stage 4
Thrust	1780 N to 17800 N	650 N to 2250 N	100 N to 1150 N	45 N to 225 N
Sea Level Specific Impulse	> 200 s	N/A	N/A	N/A
Vacuum Specific Impulse	> 255 s	> 300 s	> 300 s	> 300 s
Thrust to Weight	> 50:1	> 80:1	> 80:1	> 80:1
Burn Duration	< 120 s	> 120 s	> 120 s	> 120 s

None of the above range of engines is beyond the technological capability of even amateurs today. They merely have to be developed and made available. They need to be reliable, modular and sufficiently efficient in thrust and mass. The opportunity exists now.

An additional separate classification will be needed for spacecraft propulsion systems. These will likely be various forms of upper stage vehicle electric propulsion systems. Electric propulsion systems can be many times more efficient with their propellants than any chemical propulsion. As suggested, there will be a whole array of modular plumbing parts needed to integrate these engines into vehicles. Likely parts will be main propellant valves, regulators, check valves, pressure relief valves, tubing and high pressure composite pressure tanks. The parts listed above, rocket engines and plumbing parts, are nearly all that's required to bring about the microlauncher revolution at this point.

These parts will first be developed by hobbyists for their first stages. These are vehicles than can raise the upper stages above the atmosphere, to where vacuum engines can operate efficiently enough. As these first stage engines are perfected, the upper stage engines will be developed and then perfected. Quickly, the first microlauncher will go into an escape trajectory, never to be heard from again. But, this is the point where the revolution has started.

As these engines are perfected, traded, sold and made available, universities will see opportunities and develop and purchase them for their own missions. A few companies will form to provide these engines and launch services to the growing niche market that is the hobbyists and universities. As these parts are established, they will be designed into other science missions perhaps by government space agencies.

There will eventually come a day when Big Aerospace produces these engines for their government clients. When this happens, you'll know that the microlaunchers revolution has established itself. The revolutionary change in space access is then just around the corner.

The result of this process will be that microlaunchers will become affordable. When launch costs come down to thousands rather than millions of dollars, space missions will become widely available.

Legislation will be changed to accommodate the growing numbers of launch vehicles on escape trajectories, making it easier and easier to do, and establishing some standards. Welcome to the new microlaunchers future.

Launch Vehicle Introduction

Chapter 5 - Introduction to Rocketry

The key technology for the microlaunchers revolution is the small launch vehicle. If one is going to be involved in developing microlaunchers, one needs to understand the technology behind them in sufficient detail. In this chapter, we'll start reviewing the theory behind rocket design that is necessary for designing a microlauncher. An understanding of the problems and issues of rockets will give an appreciation for their capabilities and limitations and what is required to design them and use them effectively.

The Rocket Thrust Equation

A rocket engine produces thrust by burning a fuel and an oxidizer inside of a combustion chamber. The combustion creates hot gases which produce pressure. The pressure inside of the combustion chamber, **Pc**, is the pressure at which the propellants are combusted. Once burned and converted into gas, the propellants escape through the nozzle. The following diagram illustrates the various parts that we're interested in considering.

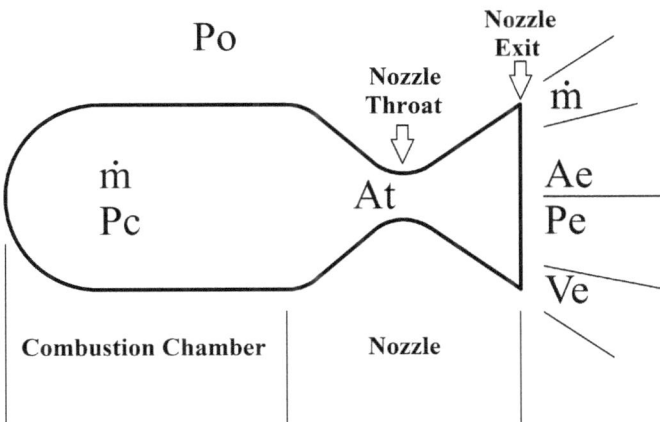

In the above image, the heavy line represents the shape of a rocket engine. The combustion chamber is to the left and the nozzle is to the right. A certain amount of fuel, defined by its mass, **m**, is burned at a certain pressure, **Pc**, inside of the combustion chamber. It produces hot gases which exit through the nozzle throat and then through the nozzle exit into the surrounding environment. The surrounding environment has a pressure, **Po**. The surrounding environment could be at sea level, for example, and then **Po** would be 101 kPa; in a vacuum, the pressure of **Po** would be near zero. The gases leave the nozzle exit at a particular pressure, **Pe**, and with a certain velocity, **Ve**. The nozzle exit has a certain area, **Ae**. If you notice in the diagram above, the mass symbol, **m**, has a dot over it. This is used to signify the mass rate through the chamber and is in units of mass per second. In other words, during continuous thrusting, the propellants are introduced into the combustion chamber at a certain rate, **m** with a dot over it, also known as **m dot**. You don't just burn 5 kg of propellant, you burn 5 kg of propellant per second; this is its m dot.

$$\dot{m} = \frac{m}{t}$$

Knowing all of this nomenclature, we can introduce the rocket thrust equation.

$$F = \dot{m} \cdot V_e + (p_e - p_o) \cdot A_e$$

This equation says that the thrust of the rocket, its force **F**, is equal to m dot times the exit velocity of the gases plus the difference in pressure at the nozzle exit times the area of the nozzle exit. There are two sub terms to this equation, the mass times velocity part and the nozzle effect part. This equation also demonstrates an important characteristic of rocket nozzles: they work better with lower pressure outside of the nozzle. In an optimally expanded rocket engine, the nozzle exit pressure, p_e, is equal to the outside ambient pressure, p_o, and the nozzle adds no extra thrust. But, as p_o approaches a vacuum pressure of zero, the term $(p_e - p_o)A_e$ is maximized. The equation also demonstrates another important characteristic of rocket nozzles, if p_e is less than p_o, then the nozzle force at that pressure subtracts from the overall thrust. This represents an important concept in rocket engines: overexpansion and underexpansion.

UNDEREXPANDED   **OPTIMALLY**   **OVEREXPANDED**
                             EXPANDED

In an underexpanded nozzle, the pressure of the gases leaving the nozzle exit, p_e, are higher than the ambient pressure, p_o, and so they continue expanding. In an optimally expanded nozzle, the pressure of the gases leaving the nozzle, p_e, equals that of the ambient pressure, p_o, and so they do not expand further. In an overexpanded nozzle, the pressure of the gases leaving the nozzle exit, p_e, is below the ambient pressure, p_o, and so they are compressed inwards. If the nozzle is greatly overexpanded, the gases will disconnect from the nozzle and can be unstable, so a greatly overexpanded condition is to be avoided. Nonetheless, it is common practice to overexpand the gases somewhat to improve the performance of the nozzle at altitude. A general rule says that one should not overexpand the gases to less pressure than about 45% of the ambient pressure, p_o, to avoid dangerous instability.

Since the atmospheric pressure decreases with altitude, one can see that rocket engines are more efficient at higher altitudes (and in a vacuum). One simple model of atmospheric pressure versus altitude is:

$$p = 101.353 \cdot e^{-0.138714 \cdot a}$$

where:
    p is the pressure in kiloPascals
    a is the altitude in kilometers

Although this doesn't accurately match the atmosphere, being off a few percent at times, it does very closely follow the trend of the atmosphere and it shows that atmospheric pressure falls off almost exponentially with altitude. Therefore, based on the rocket thrust equation, we can see that the second sub term of the rocket thrust equation increases with the exponent of altitude. Here's a simple example of I_{sp} vs. altitude for a particular rocket design.

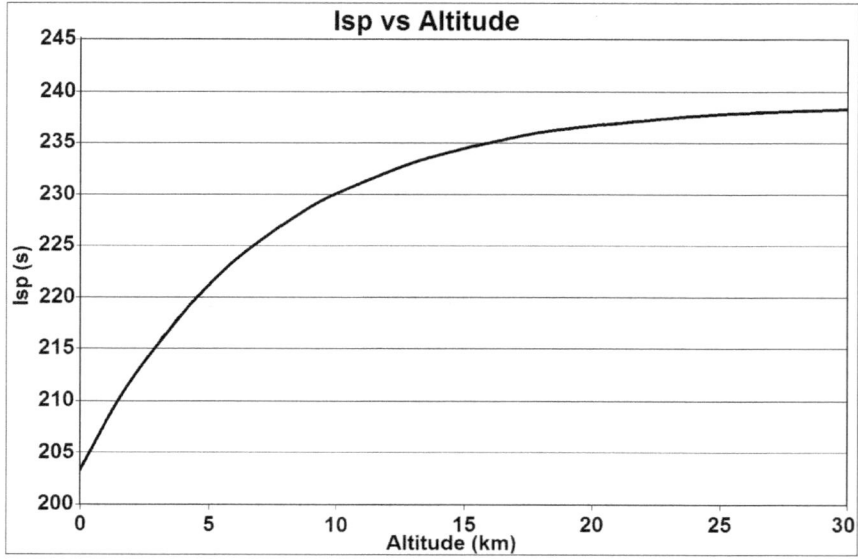

Since most of a launch vehicle's flight time is at higher altitudes, its overall average I_{sp} during flight will be closer to the vacuum value than the sea level I_{sp} value. This fact also emphasizes the need for longer burn times so that the rocket's I_{sp} can improve with longer durations at higher altitudes.

Expansion Ratio

There's another important idea that's related to rocket nozzles that we'll need to know later. It's known as the *expansion ratio* of the nozzle. The expansion ratio is the ratio of the area of the nozzle exit divided by the area of the nozzle throat.

$$\varepsilon = \frac{A_e}{A_t}$$

The expansion ratio affects the gas exit velocity, *V_e*, and the gas exit pressure, *P_e*, of the nozzle.

Specific Impulse

Earlier, in chapter 3, the idea of specific impulse was introduced, symbolized as I_{sp}. However, this is such an important concept with regards to rockets, that it is worth reviewing it in depth. Here is the full I_{sp} equation in two different forms.

$$I_{sp} = \frac{F \cdot t}{m \cdot g} = \frac{F}{\overset{\bullet}{m} \cdot g}$$

Therefore, the specific impulse equals the thrust force times the time over which that force was exerted divided by the mass of the propellant to produce that force times the gravitational acceleration. In the earlier chapter, it was stated that I_{sp} defines the efficiency of a rocket. I_{sp} defines the relationship between the amount of fuel used to produce a certain amount of thrust for a certain duration of time.

The following table illustrates some theoretical I_{sp} values for various propellants at sea level.

Propellant Combination (Pc = 1.38 MPa, Pe=101 kPa)	OF Mixture Ratio	I_{sp} (s)
Black powder (60/30/10)	1.50	90
KNO3 + Sorbitol (65/35)	1.86	110
AP + HTPB (80/20)	4.0	178
AP + HTPB + Al (70/15/15)	2.3	191
Liquid Oxygen + Ethanol (95%)	2.0	212
Liquid Oxygen + Kerosene (RP1)	3.4	216
Liquid Oxygen + Butane	3.5	221
Liquid Oxygen + Propane	3.6	221
Liquid Oxygen + Methane	3.9	226
Liquid Oxygen + Liquid Hydrogen	6.2	292

This table is only meant to allow some rough comparison between performances of a few propellants. It is not meant to imply practicality or breadth of sampling.

The Rocket Equation

One of the most important equations that needs to be known to understand rocket launchers is the Rocket Equation. You've got to figure that if it's known as *The Rocket Equation*, then it's going to be pretty important in rocketry.

In order to understand the rocket equation, we need to first be able to reference the portions of a rocket that we're interested in. Since we'll be showing them in the rocket equation, we'll give them names that facilitate their use in equations. The first parameters to understand are the various designations for component masses.

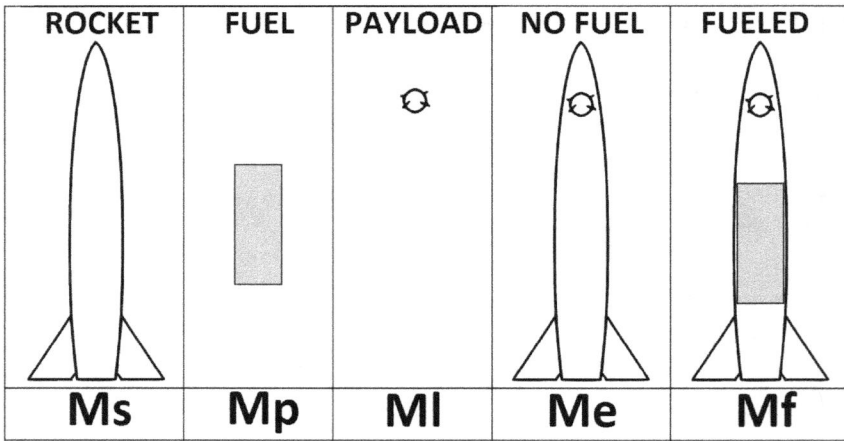

ROCKET	FUEL	PAYLOAD	NO FUEL	FUELED
Ms	**Mp**	**Ml**	**Me**	**Mf**

The above image shows the designations used for parts of the rocket. M_s is the mass of the structure of the rocket without any payload or propellant. M_p is the mass of the propellant. M_l is the mass of the payload. M_e is the mass of the rocket with the payload, consisting of M_s plus M_l. M_f is the mass of the rocket with payload and propellant. Here are the parts summarized again:

$$M_s = \text{Mass of the rocket's body structure}$$

$$M_p = \text{Mass of the rocket's propellant}$$

$$M_l = \text{Mass of the rocket's payload}$$

$$M_e = M_s + M_l$$

$$M_f = M_s + M_l + M_p$$

Given this nomenclature, the Rocket Equation states the following relationship:

$$dV = V_e \cdot \ln(\frac{M_f}{M_e})$$

What this equation says is that the maximum velocity change, **dV**, that a rocket experiences is equal to the nozzle exhaust velocity, **V_e**, times the natural logarithm of the fully loaded mass, **M_f**, divided by the propellant free mass of the rocket, **M_e**.

When dealing with rocket thrust over its flight trajectory, the integration of the rocket's thrust averages out so that we can make a useful substitution of **V_e** with the average **I_{sp}** times the acceleration of gravity, **g**. Therefore, the rocket equation is also written:

$$dV = g \cdot I_{sp} \cdot \ln(\frac{M_f}{M_e})$$

The rocket equation specifies the maximum possible velocity that a rocket can attain without aerodynamic or gravity effects. However, in a gravitational field, the rocket will be slowed down by gravity therefore this loss due to gravity must be subtracted from the expected velocity. This **gravity loss**, as it is known, is easily calculated:

$$V_{gl} = g \cdot t$$

Where, **g** is the acceleration due to gravity and **t** is the amount of time that the rocket is thrusting.

Additionally, in the atmosphere, a rocket will be slowed down by aerodynamic drag. This loss of velocity due to aerodynamics is called the **aerodynamic loss**. Aerodynamic losses are difficult to calculate; they are likely best determined by simulation along the flight trajectory. We will represent aerodynamic losses as **V_{aero}**.

One other loss experienced by a rocket is known as the **steering loss**.

This is the loss of propellant not used to accelerate the rocket, but instead used to steer it and keep it on course. We will represent it as V_{sl}. It is usually quite small and is often not included.

Altogether, we can represent the total vehicle velocity by accounting for all of the expected losses as:

$$dV_{re} = V_{act} + V_{gl} + V_{aero} + V_{sl}$$

This shows that the velocity culculated by the rocket equation, dV_{re}, equals the actual velocity attained at the end of the thrust duration, V_{act}, plus the gravity losses, V_{gl}, plus the aerodynamic losses, V_{aero}, plus the steering losses, V_{sl}. This equation can also be rearranged to illustrate the more basic fact:

$$V_{act} = dV_{re} - V_{gl} - V_{aero} - V_{sl}$$

In other words, the actual velocity seen is the velocity predicted by the rocket equation minus the gravity losses, minus the aerodynamic losses, minus the steering losses.

Rocket Equation Example

Let's look at a practical application of the equations we've introduced. Earlier we introduced three different sounding rockets: the Aerobee 150A, the Super Arcas and the Astrobee D. Let's apply the Rocket Equation to one of them to show the derivation of its theoretical velocity.

The Aerobee 150A sustainer body, its M_s, weighed 132.9 kg. Its total propellant mass, M_p, was 478.3 kg. With a payload, M_l, of 54.4 kg, the following table illustrates the mass components together.

Parameter	Value
Ms	132.9 kg
Mp	478.3 kg
Ml	54.4 kg
Me	187.3 kg
Mf	665.6 kg

The Aerobee had an average Isp of 209 seconds during its flight. Based on all of this information, we can calculate the theoretical maximum velocity, **dV**.

$$dV = g \cdot I_{sp} \cdot \ln(\frac{M_f}{M_e})$$

$$dV = 9.8 \cdot 209 \cdot \ln(\frac{665.6}{187.3})$$

$$dV = 2048.2 \cdot \ln(3.553)$$

$$dV = 2048.2 \cdot 1.2678$$

$$dV = 2597 \text{ m/s}$$

Therefore, the sustainer stage of the Aerobee 150A, with a payload of 54.4 kg had a maximum theoretical velocity of 2597 m/s. The literature reports that this vehicle also saw a combined gravity loss plus aerodynamic loss of 1012 m/s and that the actual velocity at the end of thrust was 1589 m/s. The Aerobee sustainer engine burned for about 50 seconds; knowing this, we can calculate the gravity loss.

$$V_{gl} = g \cdot t$$

$$V_{gl} = 9.8 \cdot 50$$

$$V_{gl} = 490m/s$$

We can extract the aerodynamic losses knowing this. Since aerodynamic losses plus gravity losses is known to be 1012 m/s and the gravity loss is known to be 490 m/s, the aerodynamic losses are determined to be about 522 m/s.

Implications of the Rocket Equation

It's not enough to merely know of the existence of the rocket equation, to design rockets using it, one needs to understand some of the implications of it. Here's the rocket equation again:

$$dV = V_e \cdot \ln(\frac{M_f}{M_e})$$

The first most obvious thing to see in the equation is that the **dV** is directly proportional to the **V_e**. Therefore, if you have higher **V_e**, you will have a higher theoretical **dV**. Following this reasoning, since the logarithm of a value is related to its input so that increasing its argument increases the result, the **dV** of the vehicle is logarithmically proportional to the mass ratio. Therefore, if you increase the mass ratio, you will get an increase in **dV** and a decrease of the mass ratio will result in a decrease of **dV**.

Looking at the other form of the rocket equation,

$$dV = g \cdot I_{sp} \cdot \ln(\frac{M_f}{M_e})$$

we can see that the **dV** is also directly related to the I_{sp} of the rocket engine. A higher I_{sp} will result in a larger theoretical **dV** velocity gained by the rocket vehicle.

The following graph shows the **dV** attained according to mass ratio and Isp.

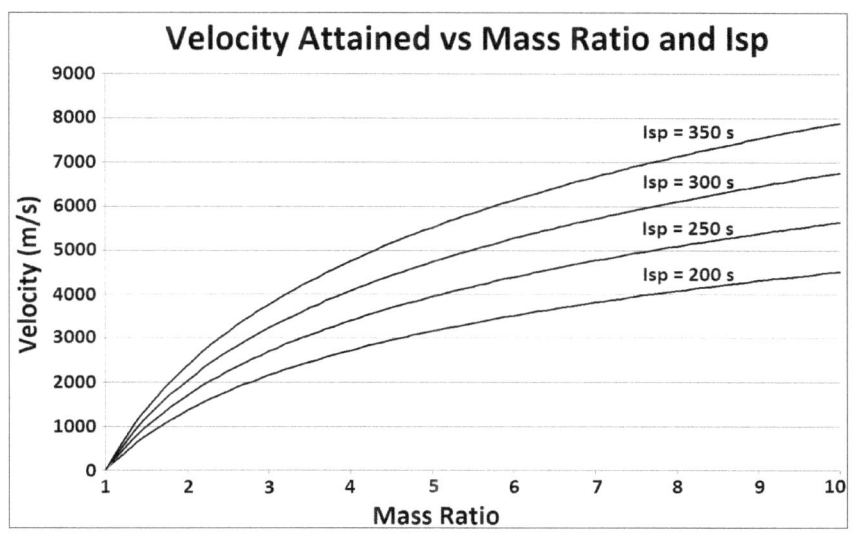

Velocity Attained vs Mass Ratio and Isp

Looking at the graph, if we wanted a rocket able to have a velocity increase of 4000 meters per second, then one must have a mass ratio of 7.70 if the Isp is 200 seconds, a mass ratio of 5.12 if the Isp is 250 seconds, a mass ratio of 3.90 if the Isp is 300 seconds and greater than 3.21 if the Isp is 350 seconds. Let's look at what this means in terms of rocket dimensions for the same payload.

The following table lists two different dV values, four different Isp's and the associated Mass Ratio and mass components for the vehicles with those characteristics.

Parameter	3000 m/s				4000 m/s				Units
Isp	200	250	300	350	200	250	300	350	s
Mass Ratio	4.62	3.40	2.77	2.40	7.70	5.12	3.90	3.21	
Ml	50	50	50	50	50	50	50	50	kg
Ms	50	50	50	50	50	50	50	50	kg
Mp	362	240	177	140	670	412	290	221	kg
Me	100	100	100	100	100	100	100	100	kg
Mf	462	340	277	240	770	512	390	321	kg

Here's a representation of these changes realized as actual rocket shapes.

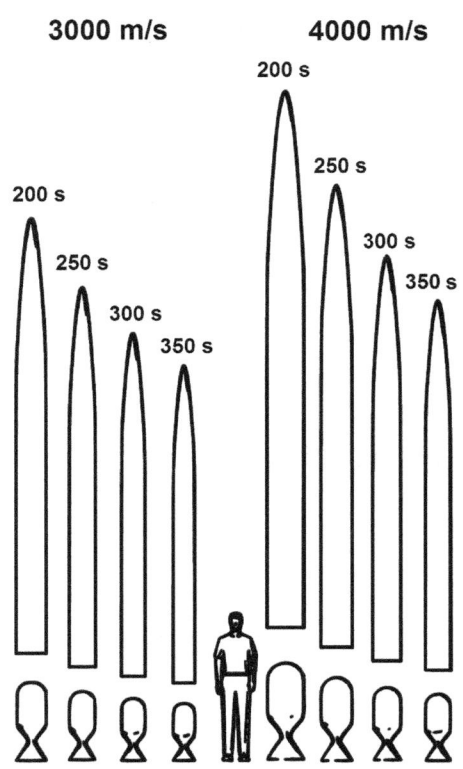

3000 m/s **4000 m/s**

200 s

250 s

200 s

250 s

300 s

350 s

300 s

350 s

There's obviously quite a great variety in sizes. To the left are rockets that have a dV of 3000 m/s and to the right are rockets with a dV of 4000 m/s. Focusing on the left side rockets, the 200 second I_{sp} rocket has to have 2.6 times the propellant as the 350 second I_{sp} rocket even though the empty rocket weighs as much as the smallest vehicle. But what if we want a vehicle that can go 4000 meters per second? The rockets have increased in size dramatically (even though they're supposed to all have the same structural mass, M_s). Some of these rockets will be reasonable to build, but some of these rockets will be very difficult to build because they're supposed to hold a very large volume of propellants but can't weigh any more than the smaller vehicles.

The important thing to realize is that there is a very non-linear relationship between the mass of a rocket and its expected performance and mass ratio. Look at the 200 second I_{sp} 3000 m/s rocket and compare it to the 200 second I_{sp} 4000 m/s rocket. The higher performance rocket weighs 66% more for only 50% more dV. This ratio gets much worse as more performance is demanded of the rocket. It's also obvious that vehicles with higher I_{sp} are significantly smaller than those with much lower I_{sp} for the same performance.

Multistage Rockets

There's one simple solution to the non-linearity of mass increase in the rocket equation. If you want to get more dV performance without significantly increasing the vehicle size, you can stage your rockets.

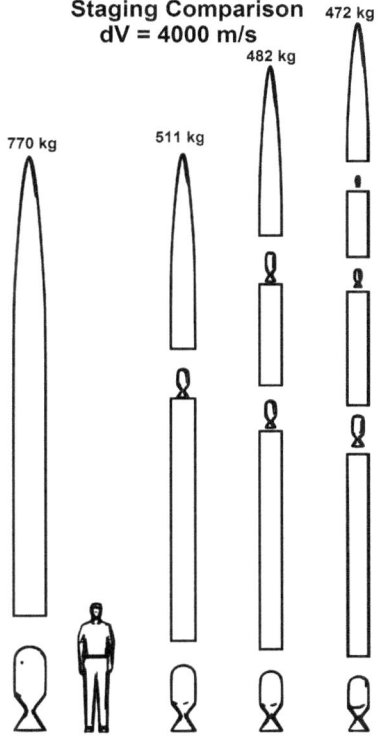

Staging Comparison
dV = 4000 m/s

Let's look at the same rocket example above and see what size vehicle we get if we try to get 4000 m/s by staging rockets. For simplicity, I'll split the dV evenly between the stages. Also, each upper stage will be the payload of the lower stage.

So the first vehicle to the left is the same 4000 m/s 200 second I_{sp} vehicle as before, but each vehicle to the right uses more stages to raise the same payload to the same dV with the same I_{sp}. As can be seen, going from 1 stage to 2 stages is a significant decrease in the mass of the vehicle. Going from 2 to 3 stages has less benefit and even less mass decrease benefit is seen as one keeps adding stages. Generally, two, three or four stages is preferred among most launch vehicles that need large speed increases, although there have been vehicles with more stages.

It should be emphasized that the masses of these vehicles in these examples is not really very reasonable (they're too light for their size), but I've wanted to emphasize the relative relationship in vehicle size using the rocket equation. In the next chapter, we'll start looking at what reasonable ranges for mass values are.

Other Important Mass Ratios

Even though the rocket equation mass ratio is one of the most important concepts to understand, it is not the only mass ratio that is spoken of in rocketry. The following table lists the various mass ratios generally used shown in the nomenclature we introduced earlier. One problem is that although these various ratios are used universally, there is no single authoritative agreement on representation of symbols. Therefore, these symbols will often be in conflict with those used in other sources.

Description	Equation
Propellant Ratio	$\dfrac{M_p}{(M_s + M_p)}$
Structural Coefficient	$\dfrac{M_s}{(M_s + M_p)}$
Payload Ratio	$\dfrac{M_l}{(M_s + M_p)}$
Rocket Equation Mass Ratio	$\dfrac{M_s + M_p + M_l}{M_s + M_l} = \dfrac{M_f}{M_e}$

It is worth familiarizing oneself with these various ratios because they will be useful in analyzing rocket designs. The Propellant Ratio and Structural Coefficient are ways of representing how sophisticated the rocket structural design is. The structural coefficient represents the ratio between the structure and the rocket with its propellant: a lower value represents a more sophisticated implementation while a higher ratio represents a less sophisticated design. The contrast to these ratios is the Propellant ratio. It represents the percentage of the full vehicle that is propellant. A more sophisticated design will have a higher Propellant Ratio than a less sophisticated design. In actuality, the Propellant Ratio is equal to 1 minus the Structural Coefficient (and the Structural Coefficient is 1 minus the Propellant Ratio). We've already discussed the implications of the Rocket Equation Mass Ratio so we won't discuss it more here.

Just to give some context for these values, let's plot some relative designs compared against each other.

A Few Select Propellant Ratios

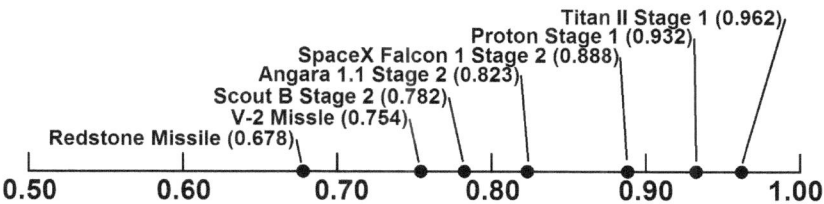

As can be seen in the selection of a few different rocket vehicles, realistic launch vehicles have propellant ratios above 0.6, preferably above 0.7 and the highest is about 0.96. In a large selection of launch vehicles, the average will be about 0.87.

Getting into Space

We've covered a little bit about the design of rockets, enough to begin applying those ideas to getting a vehicle into space. Let's look at three different aspects of being in space: a ballistic trajectory, a circular Low Earth Orbit (LEO) and an Earth escape orbit. By a generally accepted definition, "space" starts at 100 km above sea level; this is also known as the Karman line.

Ballistic Trajectory Example

We've already seen in an earlier chapter some of the analysis related to putting sounding rockets on ballistic trajectories to space. Let's calculate the parameters of a microlauncher first stage. The mission is a typical microlauncher mission to raise a payload of 10 kg to an altitude of 100 km. Let's assume the following parameters:

Parameter	Value	Units
Payload	10	kg
Target Altitude	100	km
Average Isp	245	seconds
Thrust Duration	40	seconds
Gravity Losses	393	m/s
Aerodynamic Losses	640	m/s
Propellant Ratio	80	Percent

First, let's calculate the velocity required to reach that altitude. Simple physics formulas tell us the velocity we need to attain. In the following equation, s is the altitude attained, s_0 is the starting altitude, v_0 is the starting velocity, a is the acceleration of gravity and t is the time:

$$s = s_0 + v_0 t + \frac{1}{2} a t^2$$

Using basic algebra, we can solve this to determine the time of flight, given that $s_0 = 0$ and $v_0 = 0$:

$$s = \frac{1}{2} a t^2$$

$$t = \sqrt{\frac{2s}{a}}$$

Therefore, the time of flight is:

$$t = \sqrt{\frac{2 \cdot 100000}{9.8}}$$

$$t = 142.9 \text{ seconds}$$

Then we can determine the velocity as:

$$v = at$$

$$v = 9.8 \cdot 142.9$$

$$v = 1400 m/s$$

Therefore, the velocity required is about 1400 m/s. The presumed aerodynamic losses are about 640 m/s and gravity losses are about 393 m/s. Then, we need a total velocity increase, dV, of about 2433 m/s. It should be pointed out, however, that this is only a first-order approximation of the velocity to attain this altitude. A more careful analysis would include other factors which we won't consider in this chapter. However, this value is sufficiently close to be a workable result.

We'll need to use the rocket equation to figure out how much propellant we need; however, first we need to calculate a substitution for structural mass from the propellant ratio. The propellant ratio equation is:

$$\text{Propellant Ratio} = \frac{M_p}{(M_s + M_p)}$$

Using basic algebra, we can extract out the **Ms** value:

$$M_s = M_p(\frac{1}{\text{PropellantRatio}} - 1)$$

Therefore, we can substitute this into the rocket equation in place of the structural mass in our calculations.

$$M_e = M_p(\frac{1}{\text{PropellantRatio}} - 1) + M_l$$

$$M_f = M_p + M_p(\frac{1}{\text{PropellantRatio}} - 1) + M_l$$

Going back to the Rocket Equation:

$$dV = g \cdot I_{sp} \cdot \ln(\frac{M_f}{M_e})$$

$$2433 \text{ m/s} = 9.8 \cdot 245 \cdot \ln(\frac{Mf}{Me})$$

With some basic algebra, we can calculate the Rocket Equation Mass Ratio, **MR**:

$$MR = e^{\frac{dV}{g \cdot I_{sp}}}$$

$$MR = e^{\frac{2433}{9.8 \cdot 245}}$$

$$MR = e^{1.013}$$

$$MR = 2.754$$

Now we know that we need a mass ratio of 2.754 for the vehicle. We can then substitute into the Mass Ratio equation to calculate the propellant mass, **M_p**.

$$MR = \frac{M_f}{M_e}$$

$$MR = \frac{M_p + M_p(\dfrac{1}{\text{PropellantRatio}} - 1) + M_l}{M_p(\dfrac{1}{\text{PropellantRatio}} - 1) + M_l}$$

After some algebraic manipulation, we derive the following equation:

$$M_p = \frac{M_l(1 - MR)}{(MR - 1)(\dfrac{1}{PR} - 1) - 1}$$

Yes, it's annoyingly complicated, but it allows us to calculate the propellant mass knowing the payload mass, **M_l**, and the mass ratio, **MR**. Substituting the presumed values for MR and M_l, we get:

$$M_p = 31.2 kg$$

Now we know the vehicle structure mass by further substitution:

$$M_s = M_p(\frac{1}{\text{PropellantRatio}} - 1)$$

$$M_s = 31.2(\frac{1}{0.80} - 1)$$

$$M_s = 7.8kg$$

And now we know the basic masses of the vehicle overall.

Parameter	Value	Units
M_l	10	kg
M_p	31.2	kg
M_s	7.8	kg

This shows how the rocket equation and basic physics and algebra can be used to calculate the mass parameters of a rocket with a specific performance. A rocket with the defined parameters will be able to lift its payload to about 100 km. Here's a diagram of the likely dimensions of this rocket; it's about the same size as the Astrobee D but a bit lighter.

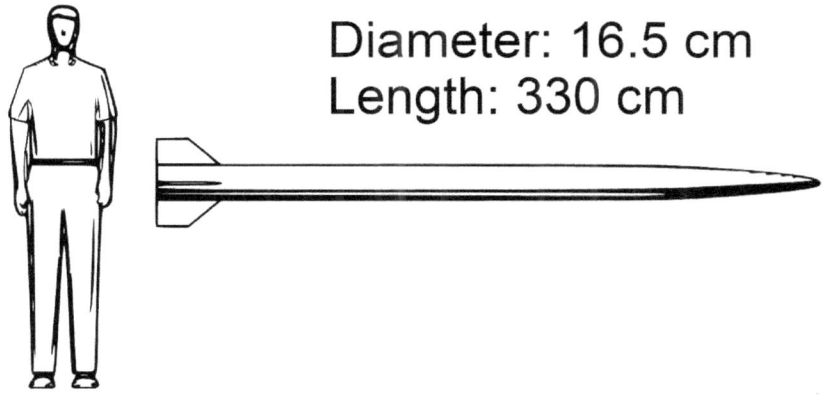

Diameter: 16.5 cm
Length: 330 cm

Circular Orbit Example

A Low Earth circular orbit is an important example of a high performance launch vehicle. Let's presume that we've used a first stage vehicle able to raise upper stages on a trajectory towards the desired altitude. Thus the two upper stages must provide the velocity to put the payload in orbit after being released from the first stage at about 65 km.

Since the upper two stages will be operating entirely in a vacuum, they can have big nozzles that provide improved I_{sp}. They also do not have to deal with any significant air drag above about 65 km altitude. Therefore, our task is to determine how much velocity these two stages must be able to provide together in order to get into orbit.

For this example, let's consider the following parameters as the basic presumptions of this exercise.

Parameter	Stage 3	Stage 2	Units
Payload	0.10	Stage 3	kg
Average Isp	310	310	s
Propellant Ratio	0.80	0.80	
dV	4.0	4.0	km/s
Target Altitude	185		km

The equation for calculating the orbital speed of a payload in a circular orbit is:

$$V = \sqrt{\frac{\mu}{r}}$$

In this equation, μ, is known as the standard gravitational parameter for the body being orbited. Since we're talking about Earth, μ has the value of 398600.4 km^3s^{-2}. In the above equation, r, represents the radius of the circular orbit. The radius will be equal to the radius of Earth, which is about 6,371.0 km, plus the altitude of the orbit. The radius of the desired orbit is 6371.0 km plus 185 km or 6556 km. Using the above orbit velocity equation, the desired orbital velocity is :

$$v = \sqrt{\frac{398600.4}{6556}}$$

$$v = 7.797 km/s$$

Without getting into any of the complications related to getting into orbit, we shall use the value of 7.797 km/s as the total **dV** goal. As can be seen from the parameter table above, we've specified significantly more dV for the two stages than is actually required. It is generally good to have some safety margin designed into a stage's requirement if it's possible.

We shall start with specifying stage 3 since it will be the payload for stage 2. Stage 3 has an I_{sp} of 310 seconds in vacuum and a propellant ratio of 0.80. Using the equations we derived in the last section, we can determine the stage masses. Reintroducing the equation to calculate the mass ratio from known parameters:

$$MR = e^{\frac{dV}{g \cdot I_{sp}}}$$

$$MR = e^{\frac{4000}{9.8 \cdot 310}}$$

$$MR = e^{1.3167}$$

$$MR = 3.731$$

Reintroducing the equation to calculate the propellant mass from known parameters:

$$M_p = \frac{M_l(1 - MR)}{(MR - 1)(\frac{1}{PR} - 1) - 1}$$

$$M_p = \frac{0.1(1 - 3.731)}{(3.731 - 1)(\frac{1}{0.8} - 1) - 1}$$

$$M_p = 0.861 kg$$

Reintroducing the equation for calculating the structural mass knowing the propellant mass:

$$M_s = M_p(\frac{1}{\Pr opellantRatio} - 1)$$

$$M_s = 0.861(\frac{1}{0.8} - 1)$$

$$M_s = 0.215 kg$$

So now we know stage 3's mass parameters.

Stage 3 Mass Parameters		
Parameter	Value	Units
M_l	0.10	kg
M_s	0.215	kg
M_p	0.861	kg
M_e	0.315	kg
M_f	1.176	kg

Now we can move on to calculating stage 2's parameters because the M_f value of stage 3 is the payload (M_l) value of Stage 2. First we shall calculate Stage 2's mass ratio:

$$MR = e^{\frac{dV}{g \cdot I_{sp}}}$$

$$MR = e^{\frac{4000}{9.8 \cdot 310}}$$

$$MR = e^{1.3167}$$

$$MR = 3.731$$

It has the same mass ratio as stage 3. Now we can calculate the propellant mass:

$$M_p = \frac{M_l(1 - MR)}{(MR - 1)(\frac{1}{PR} - 1) - 1}$$

$$M_p = \frac{1.176(1 - 3.731)}{(3.731 - 1)(\frac{1}{0.8} - 1) - 1}$$

$$M_p = 10.123 kg$$

Finally, we can calculate the structural mass of stage 2:

$$M_s = M_p \left(\frac{1}{\Pr opellantRatio} - 1 \right)$$

$$M_s = 10.123 \left(\frac{1}{0.8} - 1 \right)$$

$$M_s = 2.531 kg$$

The complete mass parameters for stage 2 are:

Stage 2 Mass Parameters		
Parameter	Value	Units
M_l	1.176	kg
M_s	2.531	Kg
M_p	10.123	kg
M_e	3.707	kg
M_f	13.830	kg

Since the two stages together have more performance than is required for the orbit, their GNC system would use its estimate of its altitude and velocity to determine when to terminate thrust and deliver the proper speed at the proper altitude.

An Escape Trajectory Example

It is desired to pursue escape trajectories with microlaunchers. Therefore, it is worth understanding some aspects of them. With an escape trajectory, specifically a parabolic trajectory, the vehicle will coast away from the central body towards a velocity of zero at infinity. The payload would not return to Earth. Escape trajectories are useful for sending payloads into deep space or to another planetary body or to an asteroid.

The velocity which produces this is defined by this equation:

$$v = \sqrt{\frac{2\mu}{r}}$$

This equation looks very similar to the circular orbit equation except for the factor of 2. This is equivalent to saying that the velocity for an escape orbit at a given radius from the center of the major body is the square root of 2 times its circular orbital velocity. We've already discussed the definition of μ in the last section.

So, if we want to place a payload on an escape trajectory we merely have to determine the velocity of a circular orbit and multiply by the square root of 2, or 1.414. From the earlier example, at 185 km altitude, the circular orbital velocity was 7.797 km/s; therefore, the escape velocity at that altitude would be 1.414 times 7.797 or 11.025 km/s. Obviously, this is a large velocity for a vehicle to attain, but it is usually attainable by adding an extra stage. For example, we could use the 310 second I_{sp} stage examples from the last section and divide the escape velocity among three stages so that each stage only has a dV of 3.675 km/s. This would require a rocket equation mass ratio of 3.352, which is a relatively low mass ratio. If we only used two stages to attain this velocity, then each stage would have a dV of 5.5125 km/s for a required mass ratio of 6.138. This is a bit more difficult to get, but it is possible. Additionally, either two upper stages or three upper stages would require a first stage able to put these stages on an appropriate trajectory.

Finally, it is worth mentioning that in attaining an escape trajectory, the rotation of the Earth imparts some velocity to the spacecraft. This value can be subtracted from the launch vehicle's overall dV requirement, thus allowing a slightly smaller vehicle.

Chapter 6 – Launch Vehicle Propellant Feed Systems

In this chapter, we'll continue into another important technical issue related to designing microlaunchers: tankage and plumbing. Plumbing such as pressure vessels, valves and tubing are inherent in most rocket propellant feed system development. Although we will do our best to cover some of the many issues that are important, there are still many issues that we cannot address in this book. It is hoped that you continue your education after this book's introduction to these topics. This section will provide enough background that you might be able to begin considering designing your own microlaunchers stages and analyzing the designs in this book.

Safety

Rocketry is an inherently dangerous activity. You should seek to familiarize yourself with all safety issues before beginning a design and attempting to build your own rockets. We will introduce some of the safety issues that you should think about, though.

From the very beginning, rockets use pressure vessels. These are pressurized gas containers which can be inherently dangerous and cause lethal pressures and shrapnel under burst conditions. Propellant tanks, themselves, are pressure vessels. The engines are pressure vessels. Pressure vessels are used to hold compressed gases for feeding into the propellant tanks. There are many safety issues associated with pressure vessels that you must familiarize yourself with.

Flammable, caustic, oxidizing and potentially explosive liquids and gases are used routinely in rockets. Some of these are very cold, a condition which introduces its own hazards. Rocketry is often experimental at its very foundation and one is working with unestablished limits on parts and materials. There are many materials with very high internal energies. Therefore, caution and safety are required in thinking about using these materials.

Despite all of these hazards, rocketry can be safe and rewarding. You should familiarize yourself with your local, state, and federal laws regarding the storage, transportation and use of pyrotechnic materials. You should also familiarize yourself with relevant standards regarding such things as pressure vessels and handling of oxidizing and flammable materials. If possible, you should join a rocketry group to learn about the safe handling, transportation and use of materials related to rocketry.

There are a few very important standards with which you should familiarize yourself. The US Department of Transportation (DOT) has standards and regulations related to rockets, propellants and pressure vessels. The American Society of Mechanical Engineers (ASME) has a number of standards that are well respected with regard to developing safe pressure vessels.

You should also familiarize yourself with the processes for establishing safety limits of pressure vessels, specifically safely establishing their burst pressures and appropriate safety factors. You should always design with safety in mind. You should always establish, document and practice safe processes and wear appropriate safety equipment.

Strength of Materials

To understand how to design safe mechanical devices, you need to learn a bit about materials, their properties and limitations. This section is a brief introduction to some of the ideas you should learn about. As you start designing with materials, you should learn about them and their properties. There are numerous excellent textbooks on this subject.

There are a few terms you should familiarize yourself with in order to think clearly about material mechanical properties. The first is **stress**. Stress is the condition experience by a material when a force is applied to it. There is compressive stress, tensile stress and shear stress. Compressive stress is related to a force which tries to squeeze an object smaller. Tensile stress is related to a force that is trying to pull an object apart. Shear stress is related to two forces operating in opposite directions on opposite sides of a plane such that they would cause the material to slide along the plane. Stress is measured in units of force divided by area; this makes it equivalent to pressure and is expressed in units of Pascal. The following diagram shows these concepts illustrated.

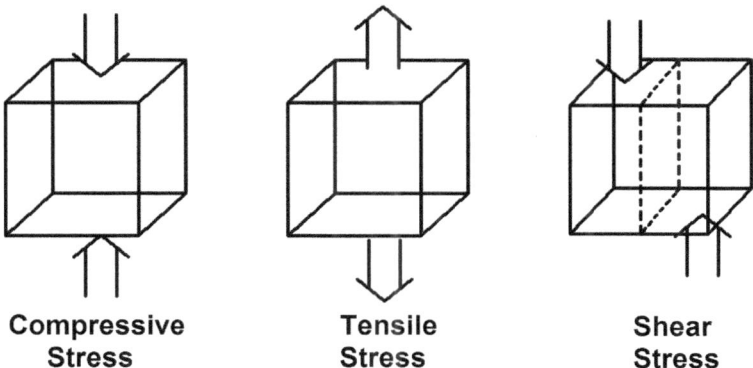

| Compressive | Tensile | Shear |
| Stress | Stress | Stress |

In this chapter, we're most interested in material properties that are significant in pressure vessel design. Therefore, you should be knowledgeable of the meaning of the term **tensile stress**. In pressure vessels, the forces of pressurized gases produce tensile stresses that we need to be highly concerned with.

Another term one should familiarize himself with is **Yield Strength**. When a force is applied to a material, it will deform basically linearly up to a point where the material no longer responds linearly. In this non-linear region, the material is beginning to enter into its failure zone. The first point where a material starts to enter into its failure zone is known as its Yield Strength. Below the yield strength, the material will generally maintain its shape and properties repeatedly. Any force that pushes a material beyond its yield strength results in permanent deformations of the material. Continued increase of these forces will ultimately result in an upper limit to its ability to contain the stresses. In tensile stresses, this failure point is known as the **Ultimate Tensile Strength** of the material. If one stresses a material further beyond its ultimate strength one will reach a final failure point known as its **Breaking Strength**. These properties change with the temperature of the material. Summarizing, a material reacts very linearly to a force applied to it, up until its yield strength beyond which the material will begin to permanently deform and beyond which it will reach its ultimate tensile strength and final failure at its breaking strength.

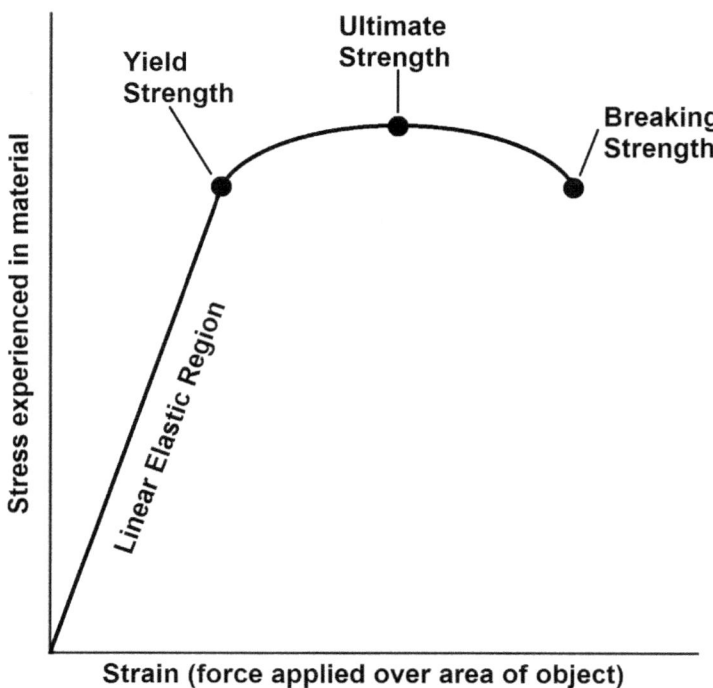

Strain (force applied over area of object)

For the most part, in designing predictable and safe devices, we wish to remain below the yield strength of the material. Therefore, when designing with materials, you should know the various relevant properties that affect safe operation. The following table shows some examples of materials, their densities, and their yield and ultimate tensile strengths at room temperature.

Material	Density (g/cc)	Yield Strength (MPa)	Ultimate Tensile Strength (MPa)
Aluminum 3004-H32	2.72	172	214
Aluminum 6061-O	2.70	55.2	124
Aluminum 6061-T4	2.70	145	241
Aluminum 6061-T6	2.70	276	310
Mild Steel 1018	7.85	370	440
A36 Steel	7.80	250	500
2800 Maraging Steel	8.00	2617	2693

As can be seen, there are a wide variety of strength properties for materials and even very similar materials can have very different strengths. Additionally, these properties can change significantly if they've been welded or heated. Finally, these properties are theoretical approximations which may never actually be met with the particular sample you're using.

Therefore, there is another term that one should be familiar with: **safety factor**. The safety factor is a multiplier that is used to ensure that one avoids the strength limit of the material. The equation is:

$$SafetyFactor = \frac{MaterialStrength}{DesignLimit}$$

Therefore, if you have a safety factor of n, you divide the strength of the material by n and design your device never to exceed that value in the specified normal operating range. For example, if you are using 6061-T6 Aluminum with a safety factor of 2, you would not allow the operating tensile stresses to exceed 276 MPa ÷ 2 or 138 MPa (presuming it hasn't been welded or heated to change the basic strength properties).

Although it is always good to allow a safety factor in your designs, it does come with a price. The safety factor is also equivalent to the extra material that is designed into your device to ensure safety. So, if you have a safety factor of 1.5, then you've got 1.5 times as much material as is theoretically needed to meet your maximum expected stresses. On rockets, this extra mass means less overall velocity performance capability. Nonetheless, it is necessary to design a safety factor into your devices to ensure that you never experience catastrophic failure.

If you don't need absolute velocity performance, you might consider using a safety factor above 2.0. For example, pressure fittings normally have a safety factor of 4 and commercial pressure vessels often have safety factors of 3.5 or more. However, in some aerospace usage, it is common to use a safety factor of 1.5 which may go as low as 1.25 in places. Of course, aerospace designs are extensively tested to ensure that the safety factor standards are actually met. Aerospace companies also precisely control their material properties to minimize the chance of failure.

Pressure Vessels

The propellant tanks used in launch vehicles are thin-walled pressure vessels. Additionally, pressure vessels are used to store pressurized gases for controlling the pressure inside the propellant tanks (and for other purposes). It could be pointed out that the rocket engines themselves are pressure vessels; therefore, pressure vessels are one of the fundamental technologies in launch vehicles. To design a microlauncher first stage, you need to understand some of the basic ideas about designing pressure vessels.

A thin-walled pressure vessel is a closed container designed to hold pressurized gases or liquids at pressures above the surrounding pressure conditions. It is considered thin walled if the wall thickness is less than $1/20^{th}$ of the smallest radius. Because their pressures are above the surrounding pressure conditions, they will generally only experience significant tensile stresses (and not significant compressive or shear stresses). In this section, we will focus on cylindrical pressure vessels because these are the kinds most often used in first stage propellant tanks.

To start with, let's consider a cylindrical tank with hemispherical endcaps. The following diagram shows one. To the left is a sectional view, in the middle is a side cutaway diagram and to the right is a top cutaway diagram.

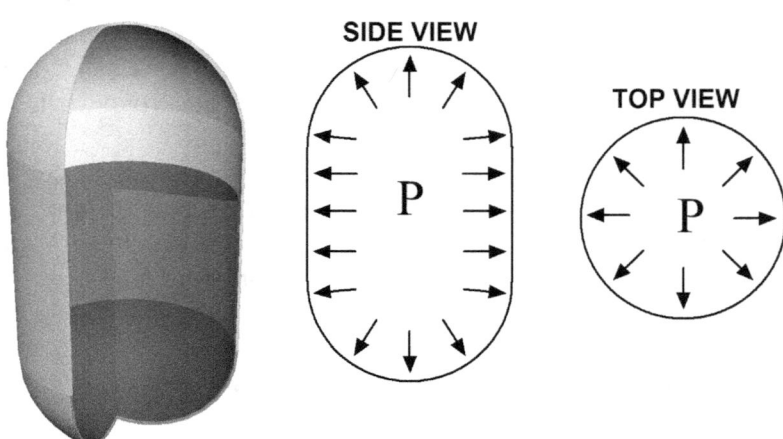

Inside of the tank is pressurized fluid at pressure, **p**. The pressure pushes outwards with a force onto the skin of the pressure vessel. The skin of the vessel distributes the forces and experiences a tensile strain as it holds in the pressure. The strain on the lengthwise skin of the pressure vessel experiences a strain that is distributed lengthwise through the cylindrical portion as well as through and around the hemispherical endcaps. The strain on the top view sectional is distributed through the skin of the pressure vessel around the circumference of the pressure vessels. These two forces, the lengthwise strain and the circumferential strain, are orthogonal to each other, or at right angles to each other, so they don't influence each other.

Engineers have identified equations that describe the magnitude of the stresses in the pressure vessel. There are three equations that are relevant to a cylindrical pressure vessel with hemispherical endcaps.

First, let's look at the easiest one which is the force impressed around the circumference of the cylinder. This is known as **hoop stress**. The equation for calculating the hoop stress is shown here.

$$\sigma_h = \frac{p \cdot r}{t}$$

Where σ_h is the hoop stress, **p** is the pressure in the tank, **r** is the radius of the tank and **t** is the thickness of the wall. The hoop stress is expressed in units of pressure (force per unit area) such as in kPa.

Second, looking at the stress through the length of the cylindrical portion, known as the **longitudinal stress**, the equation which expresses the stress is:

$$\sigma_l = \frac{p \cdot r}{2 \cdot t}$$

This equation is nearly identical to the hoop stress equation except for the 2 factor in the denominator. Therefore, under the same conditions, the longitudinal stress will be one half of the hoop stress.

Finally, looking at the stress through the hemispherical endcaps, the equation which expresses the stress is:

$$\sigma_\theta = \frac{p \cdot r}{2 \cdot t}$$

This equation, the **spherical stress**, is identical to that of the longitudinal stress and is also always half that of the hoop stress of the cylindrical portion (for endcaps with the same thickness as the cylindrical portion).

Therefore, the largest stress that a thin-walled pressure vessel experiences under pressure is developed from the hoop stress. The hoop stress is the determining factor for the overall design of this kind of pressure vessel.

Let's look at an example to emphasize these equations. Irrigation tubing is widely available in sizes useful for building microlauncher propellant tanks. It is often a high quality 3004-H32 aluminum in several thin-shelled sizes. Unlike most tubing and pipe commercially available, it comes in much thinner walled sizes than is often found from most pipe or tube suppliers. This makes it suitable for pressure vessels used in small launch vehicles.

Suppose we have an irrigation derived aluminum tube with hemispherical endcaps. The tube is 20 cm [8 inches] in diameter, 1 meter in length and the thickness is 1.30 mm [0.051 inches] for both the cylindrical and hemispherical portions. What is the maximum pressure that this pressure vessel can hold if the material is 3004-H32 Aluminum and we want a safety factor of 1.5?

Since the hoop stress is the determining factor, and the yield strength of this aluminum is 172 MPa, with a safety factor of 1.5, we can only allow 114 MPa of hoop stress in the cylinder. Thus:

$$\sigma_h = \frac{p \cdot r}{t}$$

$$114 MPa = \frac{p \cdot 10cm}{0.130cm}$$

$$p = \frac{114MPa \cdot 0.130cm}{10cm}$$

$$p = 1.482MPa$$

Therefore, to maintain the vessel within a 1.5 times safety factor of its yield strength, we should not pressurize it more than about 1.48 MPa.

Estimating the Mass of a Pressure Vessel

It is sometimes useful to be able to estimate the mass of a pressure vessel. This can be done quite easily by multiplying the area of the vessel skin by the thickness and material density. Therefore, for the cylindrical portion of a cylindrical pressure vessel, the area can be calculated as

$$Weight_{cyl} = \pi \cdot d \cdot l \cdot t \cdot \rho$$

Where **d** is the diameter, **l** is the length of the cylinder, **t** is the thickness and ρ is the material density. For the hemispherical endcaps, you can treat the two hemispheres as one sphere for the purposes of calculating the mass

$$Mass_{sphere} = 4 \cdot \pi \cdot r^2 \cdot t \cdot \rho$$

Putting these together, the mass of a cylindrical pressure vessel with hemispherical endcaps can be estimated as

$$Weight_{pv} = \pi \cdot t \cdot \rho \cdot (d \cdot l + 4 \cdot r^2)$$

Using this equation, we can estimate the mass of the pressure vessel in the earlier example. If the diameter is 20 cm, the vessel cylindrical portion's length is 1 meter (100 cm) and the thickness is 1.30 mm and it is made of 3004-H32 aluminum, the mass is approximately

$$Weight_{pv} = \pi \cdot t \cdot \rho \cdot (d \cdot l + 4 \cdot r^2)$$

$$Weight_{pv} = \pi \cdot 0.13 \cdot 2.72 \cdot (20 \cdot 100 + 4 \cdot 10^2)$$

$$Weight_{pv} = 1.111 \cdot (2000 + 400)$$

$$Weight_{pv} = 2664g$$

Therefore the mass of the vessel is about 2.7 kg. Calculated values from equations should be seen as lower bounds on the masses. Actual tanks will weigh a bit more due to such things as welds, fittings and fasteners.

Pressure Vessel Performance Factor

There is a commonly used index which allows comparison of performance between different pressure vessel technologies. It is also useful for estimating performance and mass of a rocket's propellant tanks. It is known as the pressure vessel **Performance Factor**. The equation for the performance factor is

$$PF = \frac{p \cdot v}{m}$$

Where **PF** is the performance factor, **p** is the burst pressure of the vessel (in kPa), **v** is the contained volume of the tank (in cubic meters) and **m** is the mass of the empty vessel (in kg). The result is in units of hectometers (hundreds of meters). A very strong material like carbon fiber might produce a pressure vessel having a performance factor value of 380 hectometers, or 38000 meters.

Looking at our earlier cylindrical pressure vessel, if we presume that the burst pressure was about 3.1 MPa, the volume is 37.3 liters, the thickness was about 1.3 mm and the mass is about 2.72 kg, then we can calculate the performance factor:

$$PF = \frac{p \cdot v}{wt}$$

$$PF = \frac{3100 \cdot 0.0373}{2.72}$$

$$PF = 42.5 hectometers = 4250m$$

This number is useful in several ways. First, you can use it to compare against the performance factors of other tanks and tank technologies. Second, knowing an approximate performance factor, you can estimate the mass of a tank for a given pressure and volume.

For comparative purposes, the following table lists examples of various pressure vessel technologies and their performance factors. This table is only meant to give an idea of the ranges and relative performance of some different construction techniques. It is not meant to be authoritative.

Tank Technology	Performance Factor
3004-H32 Aluminum Tank	4320 meters
2219-T87 Aluminum	10160 meters
Kevlar over 2219-T62 Aluminum	18796 meters
410 Stainless Steel	9576 meters
Titanium Lined Composite Overwrapped Vessel	38100 meters

As can be seen, it is possible to get much greater pressure vessel performance with better materials. On the other hand, these performance factors do not take into account limitations caused by flight loads; observed performance factor values can be somewhat lower. But, under ideal conditions, some composite overwrapped pressure vessels get performance factors well over 25000 meters.

The second way that you can use the performance factor is as a first-order estimate of a tank's mass knowing representative performance factors for similar vessels. For example, it might be possible to use 410 stainless steel in the earlier example and have the tanks weigh about half what they did with the 3004-H32. Of course, there are other factors that affect the performance of particular materials and designs in actual use.

Safety Testing of Pressure Vessels

Pressure vessels should be tested to establish their safety. In general, any pressurized tank should be treated as an explosive device. Any pressurized tanks have the potential to cause severe harm if compromised. Testing is done in two different ways: destructive testing and non-destructive testing.

Destructive testing is performed to ascertain whether a given construction process will produce the designed performance (usually by establishing that the actual burst pressure is near the designed burst pressure). Destructive testing should never be performed with compressible gases like air or nitrogen; it should be performed with non-compressible fluids like water. This will minimize the likelihood of hazardous explosions at high pressure. The normal process is to use a pump with a sufficient upper pressure capability to pump water into the vessel while measuring the pressure in the vessel. When it bursts, the burst pressure is noted; this establishes that the as-constructed tank meets the designed limits. Destructive testing is important to establish that such things as welds have not compromised the strength of materials in undesirable ways.

Non-destructive testing may be performed on a tank to establish that it is not leaking and that it is able to operate at the upper limit of its normal operating pressure range. Although water is the preferred medium to perform this test too, sometimes a compressible gas is used as long as the operation is performed remotely and treated as a very hazardous device. Remember, any pressurized tank can become a bomb if it is compromised. Therefore, all testing of pressurized tanks should be performed remotely. It should be filled remotely, operated remotely and pressure released remotely. The safe distances should be established with the appreciation that a tank burst could produce high speed shrapnel. Shielding should be considered where applicable.

Other Important Plumbing Components

Fittings and Tubing

Fittings and tubing are used throughout small rockets to connect propellant feed systems together. Although it may be worth considering using brazed fittings, there are many reasons to have fittings in a rocket.

There are roughly two classes of fittings that you are likely to encounter and use in rockets: NPT (or National Pipe Thread) and AN fittings (short for Army-Navy). You can usually find both types at a hardware store, but you will be better served by using automotive performance fittings or industrial versions because they're available in aluminum (and are thus lighter).

NPT threads are tapered so that they seal tighter as they are screwed in. These fittings are sized for the inside diameter of pipe and therefore the thread is larger than that. It is common to use Teflon tape to help seal these threads.

AN fittings are a flared fitting using straight threading and metal-to-metal seals. They are specialized for flexible hoses and solid tubing. AN fittings are sized from -2 to -32 where each dash number refers to the number of sixteenths of an inch in the thread. Therefore, AN-2 is 2/16" or 1/8"; AN-16 is 1". One does not use Teflon tape to seal them but there are crush washers available to facilitate sealing (though they are likely not generally needed).

There is a difference between tubing and piping. Both are often used in rockets for conveying fluids and gases. The difference is primarily related to the way that the diameter is specified. Tubing is specified by the outside diameter and the wall thickness.

Relief Valves

Relief valves are an important part in a safe propellant feed system. They release pressure beyond their specified pressure point. This ensures that should the vessel be pressurized above its maximum operating pressure, an unpleasant failure will not occur.

Pressure Regulators

Pressure regulators reduce a higher pressure down to a lower pressure. They are often used to regulate the gas supplied from a high pressure tank (say at 20 MPa) down to a workable, safe pressure used in propellant tanks (say at 1 MPa). Pressure regulators have a limited specified rate of feed, designated by a flow capacity rating called *Cv*.

You should familiarize yourself with this specification to ensure you have adequate flow.

Orifices

Orifices are used to regulate the rate of flow of gases and liquids. They are very important for ensuring that fluids and gases mix at the desired rates.

Burst Disks

Burst disks are a one-time use valve mechanism that opens once a certain pressure is reached. They generally have two uses: as a safety mechanism and as conditioning mechanism. As a safety mechanism, they are similar to relief valves except that they are one-time usage. As a conditioning mechanism, they only allow flow once a specified pressure has been attained. They are sometimes preferred over valves in places because they have zero leakage until the operating pressure is reached.

Valves

There are many different kinds of valves used in rocketry. Some valves release the gases that pressurize the propellant system, some valves release and control the flow of propellants into the rocket engine. In rocketry, since it is necessary to operate valves under computer control or from a distance, there are many kinds of motorized or servo controlled valves used.

Check Valves

Check valves are an important specialized type of valve that only allows the flow of gases or liquids in one direction. This ensures that there is no undesirable flow, say from a fuel tank into the oxidizer tank.

Tank System Design

A complete propellant feed system will include one or more propellant tanks plus a number of fittings, valves and hoses to convey the useful fluid. There will generally be a pressurant either kept in a separate pressurant tank or maintained in the propellant tanks.

There are standardized symbols used for each of the components in a tank system. The American National Standards Institute (ANSI) and the International Standards Organization (ISO) provide well respected design symbol systems. Using these symbol systems facilitates tank system design and communicates the ideas about tank designs effectively.

The following diagram shows a typical schematic for a complete propellant system.

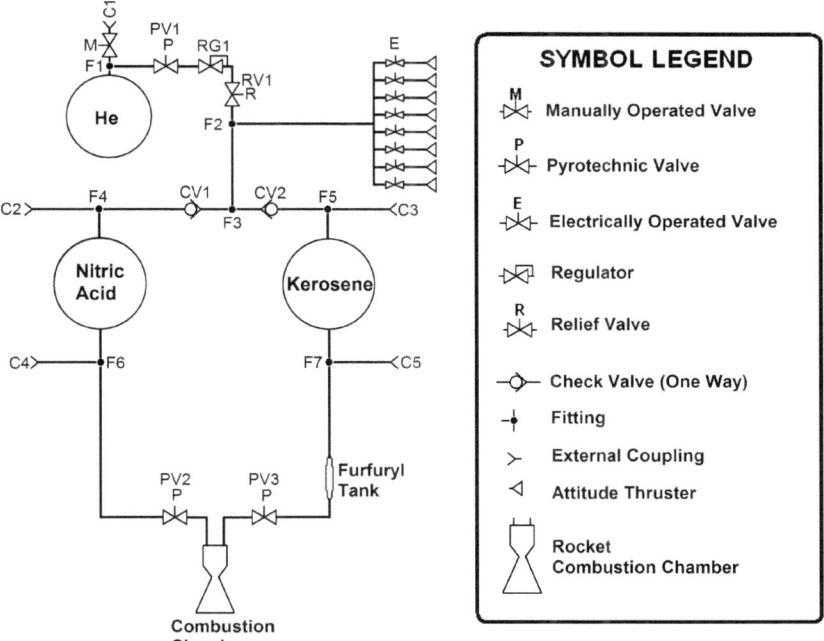

As can be seen, there is a pressurant tank, valves to control the flow, a regulator, a relief valve, check valves to control the flow direction, tanks, fittings and couplings for external access to fill and drain the tanks. The lines connecting these components are the various tubes used to connect everything. This is representative of the kind of diagrams used for designing propellant feed systems.

Chapter 7 - Rocket Engine Design

Liquid rocket engines are the most likely and best way to implement microlaunchers. Solid propellants have some benefits, but it's harder to make them work at the lower thrust levels needed with smaller rockets while keeping the rocket acceleration low. In this chapter, we'll look at the design equations associated with liquid rocket engines. This will provide necessary insight into the details of engine operation in later chapters.

Rocket Combustion

We've introduced the rocket engine in concept, but let's look at it with one more additional detail: the injector. The purpose of the injector is to atomize the propellants suitably for combustion. It is usually found at the head of the combustion chamber opposite the nozzle. The propellants are introduced into the injector at a pressure significantly above the chamber pressure, allowing a sufficiently high flow rate of the propellants.

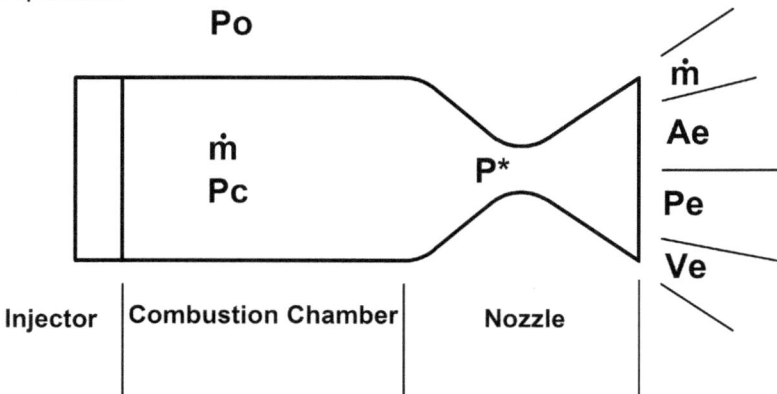

Once the propellants leave the injector and enter into the combustion chamber, they undergo a combustion process which converts their chemical energy into thermal energy.

High amounts of heat are released and the propellants chemically react to produce byproducts of that reaction. Because of the heat, these product gases expand and are propelled towards the nozzle throat as the escaping gases lower the pressure there.

The Design Process

Designing a rocket engine follows pretty closely from the combustion process itself.

The first step is the selection and introduction of propellants into the combustion process. This is usually specified as an oxidizer and a fuel with a certain mixture ratio at a certain pressure and at a certain mass flow rate (mdot). The mass flow rate at this stage is estimated from the expected engine I_{sp}.

The second step is the estimation of the combustion byproducts. Normally, the combustion products are too complicated to deal with in their precise form, so they are conceptually visualized as an ideal gas which is much easier to deal with mathematically. Therefore, after (presumed perfect) combustion, the resulting gases can be seen as a single ideal gas with uniform properties and behavior. The way that this ideal gas is described is in terms of several properties: its temperature, what is known as its "gamma" and its molecular weight (or gas molecular weight, GMW). These three properties are what are needed to be able to understand how the hot gases will behave as they go through the nozzle.

The third step is to estimate the properties of the ideal gas as it flows through the nozzle. Therefore, we estimate the temperature and pressure of the gas as it flows through the nozzle throat.

The fourth step is to determine the area of the throat given the various gas properties as it passes through the throat.

The fifth step is to calculate the dimensions of the combustion chamber. From one more parameter, known as the L*, we can determine the volume and the dimensions of the combustion chamber.

The final sixth step, knowing the throat area, is to determine the expansion ratio and the nozzle exit area.

At this point, we've described the geometry of the combustion chamber and nozzle. This is what we need to know how to construct the rocket engine.

Step 1 - Selection of Propellants

When we select propellants, we can quickly determine the properties of those propellants. Such properties as their temperature and densities are known due to the nature of the particular propellants. However, at this stage, we should also select the mixture ratio of the propellants. This is usually known as the *OF Ratio* (or oxidizer-fuel mixture ratio). There are a number of sources to help you select your propellants and the proper mixture ratio.

In Appendix A of this book, we've placed some useful propellant tables to aid you in your selection. Although the tables we've provided are only for selected propellants, you can quickly determine similar information for other propellants. Looking at Appendix A, we see that there are tables for several different propellants at different pressures. We can use these tables to select pressures and mixture ratios that produce performance metrics we're interested in. There are numerous sources for mixture ratios and performances for other propellant combinations from other sources like the internet and engineering publications.

One other major source for propellant information is known as Combustion Analysis software. There are several free and useful programs that will aid you in your selection of propellants. Rocket Propellant Analysis (RPA) is an excellent quality program that provides easy analysis of various propellants. Propep, an older free program that runs on older versions of Windows and on Linux, is well-known among rocket developers. CEA (Chemical Equilibrium with Applications) is a NASA-developed program which is also available for free. Each of these programs allows the selection of different propellants, mixture ratios and pressures and can help you determine the best propellant combinations and mixture ratios.

Step 2 - Estimating the Properties of Combustion Byproducts

Once we've selected propellants, combustion pressure and mixture ratios, we can then begin estimating the properties of the combustion byproducts.

Again, the software listed in the earlier paragraph is extremely useful in estimating these properties. What we seek at this point are the combustion temperature, the gas gamma property and the gas molecular weight. These parameters are needed to determine the dimensions of a rocket chamber for a specific thrust level. The following image shows the output of the RPA program and identifies the parameters we're interested in:

```
----------------------------------------------------
Combustion parameters:
----------------------------------------------------
      Temperature:     3214.95000            K
         Pressure:        1.37900           MPa
          Enthalpy:     -62205.877          J/mol
                        -2640.735           kJ/kg
           Entropy:       277.484       J/(mol K)
                          11.780        kJ/(kg K)
     d_lnV_d_lnT:        1.7772110
     d_lnV_d_lnp:       -1.0392314
               M:       23.5562754
       Molar mass:       0.0235563         kg/mol
          Cp (eq):       179.77414      J/(mol K)
                          7.63169       kJ/(kg K)
          Cv (eq):       154.50442      J/(mol K)
                          6.55895       kJ/(kg K)
           gamma:        1.1635534|
               k:        1.1196287
               R:         352.96208      J/(kg K)
               a:        1127.16672        m/s
             rho:           1.21524       kg/m^3
            C-ph:         0.0000000  Condensed phase mass fraction
```

From this report, we can determine that Tc, the combustion temperature, is about 3215 K, the GMW=23.56 and the gas gamma = 1.164 at a *Pc* of 1.379 MPa.

Combustion programs such as RPA, Propep and CEA are extremely useful sources for this information plus much more useful information such as the expected specific impulse performance.

An alternate source for these parameters is numerous rocket engineering publications. To simplify the process, we've included tables for select propellants in Appendix A which allow you to determine the combustion temperature (*Tc*), gas gamma (*γ*) and gas molecular weight (*GMW*) for different pressures and mixture ratios.

Step 3 - Calculating Nozzle Throat Flow Properties

Now that we know the gas combustion properties, we can calculate the temperature and pressure of the gases flowing through the nozzle throat.

The temperature of the gases flowing through the nozzle is described by this equation:

$$T_t = T_c \left[\frac{2}{\gamma + 1} \right]$$

In the above equation, T_t is the throat gas temperature, T_c is the combustion temperature and γ is the gas gamma. The values of T_c and γ are derived from the last step.

The pressure of the gases flowing through the nozzle is described by this equation:

$$P_t = P_c \left[\frac{2}{\gamma + 1} \right]^{\frac{\gamma}{\gamma - 1}}$$

In this equation, P_t is the throat gas pressure, P_c is the combustion pressure and γ is the gas gamma. Again, P_c and γ are derived from step 2 earlier.

Step 4 - Calculating the Nozzle Throat Area

The following equation allows us to calculate the throat area:

$$A_t = \frac{\dot{m}}{P_t} \sqrt{\frac{R \cdot T_t}{GMW \cdot \gamma}}$$

Where A_t is the throat area (in units of meters), mdot is the mass of the propellants flowing through the chamber each second (in kg/s), P_t is the throat pressure (in Pa) calculated in step 3, R is the Universal Gas Constant (having a value of 8.31446 J/mol-K), T_t is the gas temperature at the throat in Kelvin (calculated in step 3), GMW is the gas molecular weight from step 2 (in units of kilograms/mole), and y is the gas gamma from step 2.

You might question where the value for **mdot** came from. This came from an estimate of the engine I_{sp}. The easiest source of the I_{sp} estimate is from combustion programs like RPA, Propep or CEA. Reviewing the I_{sp} equation:

$$I_{sp} = \frac{F \cdot t}{m \cdot g}$$

and applying a little bit of algebra, we get:

$$\dot{m} = \frac{m}{t} = \frac{F}{I_{sp} \cdot g}$$

Therefore, knowing the desired thrust and an estimated I_{sp}, we can estimate **mdot**.

Step 5 - Determining the Combustion Chamber Dimensions

Once we know the throat area, we can begin calculating the dimensions of the combustion chamber. Before we do that, though, we need to understand a little bit about propellant dwell time and **L***. Although we didn't mention it in the earlier part of this chapter, combustion doesn't occur immediately. It takes time for the propellants to mix and burn and turn into that ideal gas that we imagined it to be; the time for full combustion to occur is known as the chamber **dwell time**. The required dwell time is highly dependent upon the quality of the injector and its ability to properly atomize the gases. In order to ensure adequate dwell time, engineers have devised a metric known as **L*** which is used to design a chamber with sufficient volume for good combustion.

$$L^* = \frac{V_c}{A_t}$$

From the **L*** equation, we can derive the equation for the chamber volume given the throat area and the **L***:

$$V_c = L * \cdot A_t$$

By definition, **L*** is a multiplier against the throat area which ensures enough dwell time for complete combustion. It is measured in units of length such as meters. Therefore, you will hear about a rocket engine with an **L*** of 1.5 meters, for example. When you multiply the 1.5 meters of L* by the throat area, in units of square meters, you get a volume in cubic meters. This volume is the desired volume of the combustion chamber.

Unfortunately, L* must be determined experimentally, but you can use the results from others to give you an idea of what you might design to. The following table gives some representative **L*** values for various propellants.

Propellant Combination	L* (meters)
H2O2 + RP-1	1.5-1.8
Nitric Acid + Furfuryl Alcohol	1.2
Liquid Oxygen + Propane	1.0
Liquid Oxygen + Kerosene	1.0-1.3

If you choose a propellant combination and can't find a representative **L*** in the literature, a value of 1.5 meters is a good place to start. However, it is possible to reduce the **L*** if you can get exceptionally good propellant mixing characteristics.

Knowing **L*** and **A_t**, we can calculate the volume of the combustion chamber. You should make the area of the combustion chamber cross section 10 or more times that of the throat area (about 3 times larger by diameter). This is known as the **contraction ratio**. The contraction ratio is the amount that the nozzle contracts from the combustion chamber area to the nozzle throat area.

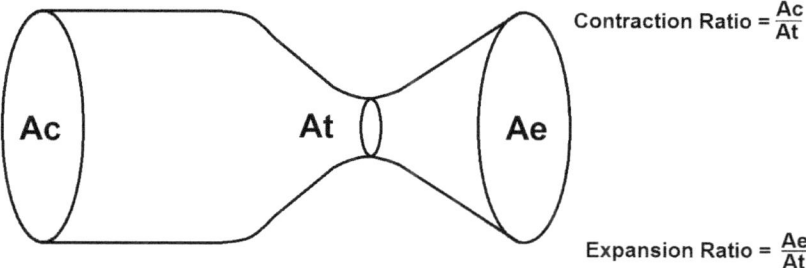

Contraction Ratio = $\frac{Ac}{At}$

Expansion Ratio = $\frac{Ae}{At}$

Step 6 - Determining the Expansion Ratio

The one last factor we have to determine is the **expansion ratio**, **ε**, of the nozzle. In order to do that, we need to determine whether the nozzle will be operating in the atmosphere or not. If it is operating in the atmosphere, then you don't want to the nozzle exit pressure to be too much lower than the pressure in which it is operating. This causes what is known as flow separation. Flow separation is the condition where the nozzle exit pressure is significantly enough below the surrounding pressure that the external atmosphere is sucked back into the nozzle. This occurs when the nozzle exit pressure is below about 45% of the external pressure and this can cause severe instabilities. If the nozzle is operating in a vacuum, then the limiting factor is the physical size and mass of the nozzle because flow separation won't occur. Vacuum nozzles tend to have expansion ratios between 30 and 150.

In either case, if you determine the nozzle exit pressure you desire, then you can determine the expansion ratio. First, select a desired exit pressure, **Pe** and know your chamber pressure **Pc** and gas gamma, **γ**. The following equation allows you to calculate the expansion ratio for your selected values.

$$\varepsilon = \frac{A_e}{A_t} = \frac{\left(\dfrac{2}{\gamma+1}\right)^{\frac{1}{\gamma-1}} \left(\dfrac{P_c}{P_e}\right)^{\frac{1}{\gamma}}}{\sqrt{\dfrac{\gamma+1}{\gamma-1}\left[1-\left(\dfrac{P_e}{P_c}\right)^{\frac{\gamma-1}{\gamma}}\right]}}$$

It is a somewhat complicated equation, but using computers and spreadsheets, this kind of equation can be automatically calculated. Another important equation is to be able to calculate the nozzle exit velocity, V_e.

$$V_e = \sqrt{\frac{2\gamma}{\gamma-1}\frac{R \cdot T_c}{GMW}\left[1-\left(\frac{P_e}{P_c}\right)^{\frac{\gamma-1}{\gamma}}\right]}$$

From this, we can calculate the instantaneous I_{sp}, the I_{sp} of the rocket engine under its specific operating conditions, as well.

$$I_{sp} = \frac{F_{thrust}}{\dot{m} \cdot g} = \frac{\dot{m} \cdot V_e + (p_e - p_o) \cdot A_e}{\dot{m} \cdot g}$$

With regard to rocket design, there's one more important thing to specify and that's the angle of the contraction and the angle of the expansion. Commonly, the contraction angle is 30° whereas the expansion angle is usually no more than 15°. This diagram shows these parameters.

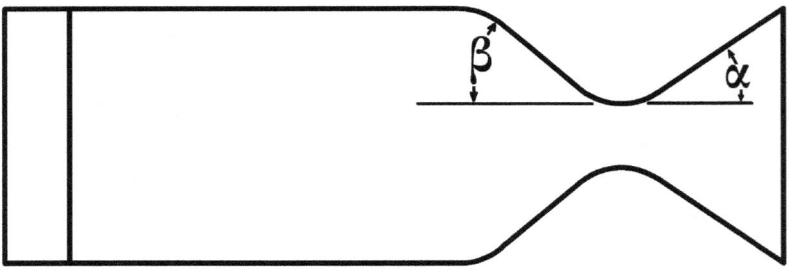

The result of all of these equations is that we are able to specify the dimensions of the combustion chamber. At this point we have all of the physical dimensions of the combustion chamber. Knowing the areas of the combustion chamber, the throat and the nozzle exit, we can calculate the associated diameters.

$$Diameter = \sqrt{\frac{4 \cdot Area}{\pi}}$$

The chamber length comes from simply knowing how to calculate the volume of a cylinder (for which we calculated with L*). Ideally, one should also consider the volume of the conical contraction region, but for simplification, we'll only calculate it as a cylinder. Doing this actually gives a slightly larger L* because there's more combustion volume.

$$Length = \frac{4 \cdot volume}{\pi \cdot diameter^2}$$

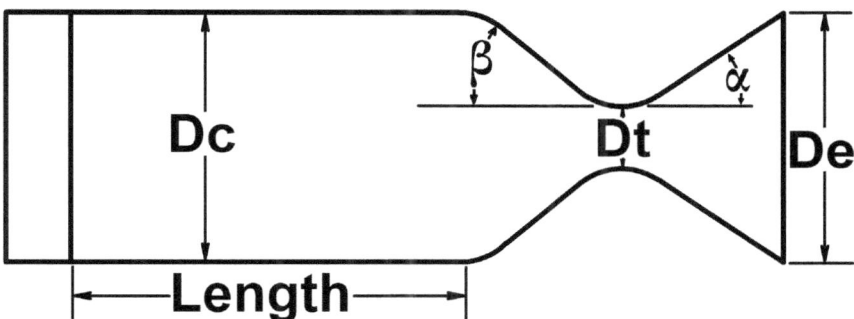

Rocket Engine Example

Let's try an example with these equations. The goal is to determine the dimensions for a rocket engine which produces 1335 Newtons of thrust, uses liquid oxygen and propane at a mixture ratio of 2.0, has 1.1 MPa chamber pressure and has an exit pressure of 101 kPa.

Using Appendix A, we can extract the T_c, gamma, GMW and the I_{sp}:

Parameter	Value	Units
Tc	2950	K
Gamma	1.17	
GMW	18.6	g/mol
Isp	220	s

Using the V_e equation, we can calculate the expected V_e.

$$V_e = \sqrt{\frac{2\gamma}{\gamma-1}\frac{RT_c}{GMW}\left[1-\left(\frac{P_e}{P_c}\right)^{\frac{\gamma-1}{\gamma}}\right]}$$

$$V_e = \sqrt{\frac{2\cdot 1.17}{1.17-1}\cdot\frac{8.31446\cdot 2950}{0.0186}\left[1-\left(\frac{0.101}{1.1}\right)^{\frac{1.17-1}{1.17}}\right]}$$

$$V_e = \sqrt{13.765\cdot 1318691\left[0.29317\right]}$$

$$V_e = \sqrt{5321558}$$

$$V_e = 2307 m/s$$

Knowing the thrust and I$_{sp}$, we can calculate the expected **mdot**.

$$\dot{m} = \frac{F_{thrust}}{I_{sp}\cdot g}$$

$$\dot{m} = \frac{1335N}{220s\cdot 9.8m/s^2}$$

$$\dot{m} = 0.619 kg/s$$

Thus, the propellant mass flow rate is 0.619 kg per second.

We now can calculate the throat pressure and temperature:

$$T_t = T_c\left[\frac{2}{\gamma+1}\right]$$

$$T_t = 2950K \left[\frac{2}{1.17+1} \right]$$

$$T_t = 2719K$$

$$P_t = P_c \left[\frac{2}{\gamma+1} \right]^{\frac{\gamma}{\gamma-1}}$$

$$P_t = 1.1MPa \left[\frac{2}{1.17+1} \right]^{\frac{1.17}{1.17-1}}$$

$$P_t = 1.1MPa \left[0.9217 \right]^{6.8824}$$

$$P_t = 0.628MPa$$

Knowing what we now know, we can calculate the throat area.

$$A_t = \frac{\dot{m}}{P_t} \sqrt{\frac{R \cdot T_t}{GMW \cdot \gamma}}$$

$$A_t = \frac{0.619}{628000} \sqrt{\frac{8.31446 \cdot 2719}{0.0186 \cdot 1.17}}$$

$$A_t = 9.857 \cdot 10^{-7} \cdot \sqrt{1038830}$$

$$A_t = 1.005 \cdot 10^{-3} m^2 = 10.05 cm^2$$

With an L* of 1 meter (100 cm), the combustion chamber volume is 100 cm * 10.05 cm^2 = 1005 cm^3.

Calculating the throat diameter, we can assign a combustion chamber diameter:

$$Diameter = \sqrt{\frac{4 \cdot Area}{\pi}}$$

$$Diameter_{throat} = \sqrt{\frac{4 \cdot 10.05cm}{3.14159}}$$

$$Diameter_{throat} = 3.577cm$$

If we let the chamber diameter be 3 times the throat diameter, then

$$Diameter_{chamber} = 3 \cdot 3.577cm$$

$$Diameter_{chamber} = 10.731cm$$

And we can calculate the length of the chamber now

$$Length = \frac{4 \cdot volume}{\pi \cdot diameter^2}$$

$$Length = \frac{4 \cdot 1005cm^3}{\pi \cdot (10.731cm)^2}$$

$$Length = 11.1cm$$

The last thing to calculate is the nozzle exit area. It is easiest to calculate the expansion ratio and then multiply the area of the throat by the expansion ratio. From this we can calculate the nozzle exit diameter. This follows from this equation:

$$\varepsilon = \frac{A_e}{A_t} = \frac{\left(\frac{2}{\gamma+1}\right)^{\frac{1}{\gamma-1}} \left(\frac{P_c}{P_e}\right)^{\frac{1}{\gamma}}}{\sqrt{\frac{\gamma+1}{\gamma-1} \left[1 - \left(\frac{P_e}{P_c}\right)^{\frac{\gamma-1}{\gamma}} \right]}}$$

$$\varepsilon = \frac{A_e}{A_t} = \frac{\left(\dfrac{2}{1.17+1}\right)^{\frac{1}{1.17-1}}\left(\dfrac{1.1}{0.101}\right)^{\frac{1}{1.17}}}{\sqrt{\dfrac{1.17+1}{1.17-1}\left[1-\left(\dfrac{0.101}{1.1}\right)^{\frac{1.17-1}{1.17}}\right]}}$$

$$\varepsilon = \frac{A_e}{A_t} = \frac{4.764}{1.934} = 2.463$$

Therefore, the expansion ratio is 2.463 and the nozzle exit area is 2.463 * 10.05 cm² = 24.753 cm². From which we can calculate the diameter.

$$Diameter_{nozzle\_exit} = \sqrt{\frac{4 \cdot Area_{nozzle\_exit}}{\pi}}$$

$$Diameter_{nozzle\_exit} = \sqrt{\frac{4 \cdot 24.753 cm^2}{\pi}}$$

$$Diameter_{nozzle\_exit} = 5.614 cm$$

The recommended values for the contraction angle is 30° and for the expansion angle is 15°. We can now have enough to make a drawing of the combustion chamber.

The following diagram shows all of the calculated values in place on a generic rocket shape, so it doesn't show the actual geometric relationships.

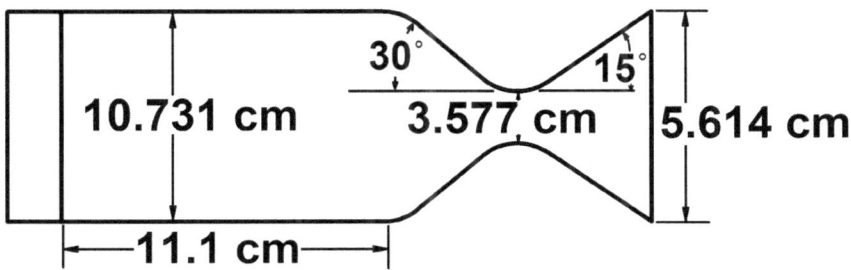

Rocket Engine Cooling

Rocket engines produce a lot of heat, more heat than can be contained without some form of cooling. A rocket engine combustion chamber would quickly heat up until it is structurally weakened in seconds if some kind of coolant isn't used. Heat fluxes can be as high as 16 kW/cm^2 in rocket engines like the Space Shuttle Main Engine. A small rocket engine like that called for in a microlauncher first stage might see an average of about 490 watts/cm^2. Even with a small rocket engine, with a combustion chamber surface area of as little as 300 cm^2, there would be total heat energy of 147 kW. Since 4.186 kJ can heat 1 liter of water one degree, it can be seen that substantial continuous heat is produced and must be removed from the combustion chamber. The small engine just mentioned would heat up 1 liter of water to boiling in about two seconds.

There are several techniques used to deal with this heat energy. It is possible to use an overabundance of chamber material to absorb the heat for short durations; these **uncooled** chambers are often used in test engines. This method is not practical for flight engines.

Ablative combustion chambers are made from materials which chemically absorb the heat as they burn, forming a protective barrier which then vaporizes and releases more ablative material underneath. The process is repeated as new surface material is exposed. Although these chambers change dimensions as they burn, the rate of ablation can be acceptable at lower pressures.

There are a few materials which can handle extreme temperatures. Known as refractory materials, they can survive the temperatures of combustion under certain conditions. Materials such as iridium, rhenium, molybdenum and some ceramics can handle very high temperatures and maintain their strength. These chambers are cooled by **radiation** of the heat into the surrounding environment. Even they are unlikely to directly survive the heat of an optimum fuel mixture, so the fuel mixtures tend to be fuel rich. Additionally, some degree of film cooling is often still used with radiation cooled nozzles.

One of the more sophisticated techniques is to run propellant in a cooling jacket around the combustion chamber. This is known as *regenerative* cooling because the heat that is removed in cooling the chamber is added to the propellant and thus enhances combustion. A variant of this idea is to simply dump the heated propellant overboard; this is known as **dump** cooling.

Film cooling relies on releasing propellant on the interior walls of the combustion chamber. As the propellant evaporates, it cools the walls as well as creates a propellant-rich zone which is cooler than the main combustion gases. Typically, there is a budgeted amount of fuel used with a well-designed film cooled combustion chamber using about 5% of the overall fuel.

Rocket Engine Cycles

An engine cycle describes the means used to pressurize the propellants into the combustion chamber.

Pressure Fed Cycles

We've already introduced the simplest cycle: the ***pressure fed cycle***. Just for review purposes, the pressure fed cycle uses a pressurized gas and tanks able to hold pressures significantly above the combustion chamber pressure to feed the propellants into the combustion chamber. *Pressure Tank* pressure fed rockets use a separate pressurized tank to feed the pressurized gas needed to pressurize the propellants.

Autogenous pressure fed rockets use the propellants themselves to generate the pressurized gas to feed the propellants into the tank. Heat is added to the propellants to produce sufficient pressure.

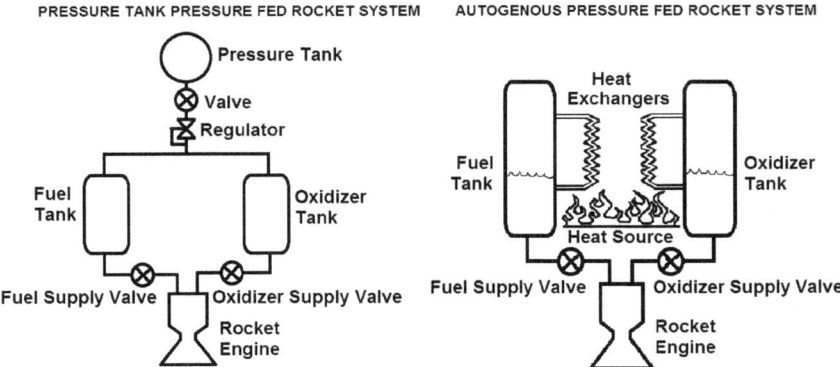

Another variation of the pressure fed cycle is the **blowdown** pressure fed rocket. A blowdown pressure fed rocket merely makes the propellant tanks larger so that the pressurized gas is stored in the same tanks as the propellants.

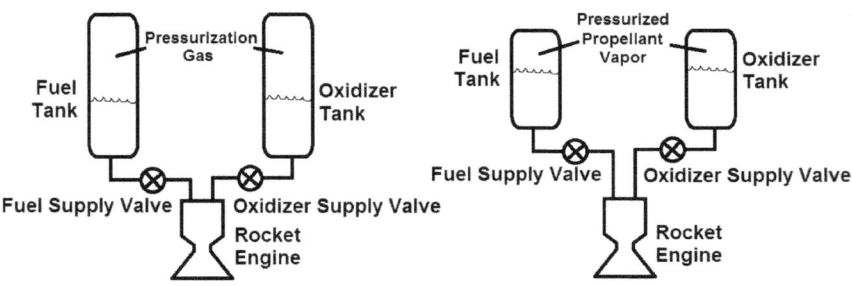

A **Self Pressurized** pressure fed rocket cycle depends on properties of the propellant to produce a pressurization gas without any additional equipment. Some propellants, like propane, have a sufficient vapor pressure that they can generate usable vaporization gas. With proper propellant conditioning before launch, it may be possible to use this property to produce sufficient pressurization for pressure feeding the propellants into the combustion chamber.

So, even though the pressure fed cycle is one of the simplest rocket feed cycles, there are several variations each with their own benefits and limitations. In all cases, as the propellant is expelled, the pressurization gases expand and cool down. Thus, available pressure can be reduced during usage. It takes careful design to get the desired performance. An additional problem is that high pressure propellant tanks can weigh significant amounts, thus lowering performance compared to other approaches.

Pump Fed Cycles

Although pumps add complexity and create opportunities for failure that don't exist with pressure-fed rockets, they do have advantages that make them worth using. With pumps, one can have much higher chamber pressure than pressure fed rocket engines and therefore higher I_{sp}. The propellant tanks can be at low pressure and the pumps can increase the combustion chamber pressure to many times atmospheric pressure.

Pump fed systems are generally turbopump systems whereby the pump is driven by a gas turbine. However, it is also possible to drive pumps with electric motors and piston engines. Improvements in motor and motor control technologies are creating opportunities to use electric motor pump systems. Nonetheless, turbine driven pump systems still allow significantly more power at this point.

Gas Generator Cycle

The gas generator is the simplest of the turbopump cycles. It is considered an "open" cycle because the gas used to drive the turbine is thrown overboard instead of going back into the rocket engine. Doing this lowers the overall performance of gas generator cycle engines a bit.

As seen in the following diagram, the gas generator takes pressurized propellants, burns them and generates hot gases that drive the turbine; after going through the turbine, the gases are released overboard. The turbine drives the pumps to provide pressurized propellants to the combustion chamber. The gas generator is usually run very fuel rich in order to keep the turbine gases at a low enough temperature not to melt the turbine. Because of this, the turbine exhaust has a lot of unburned fuel in it.

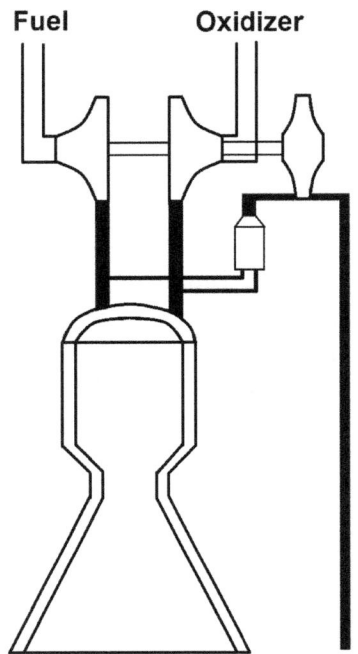

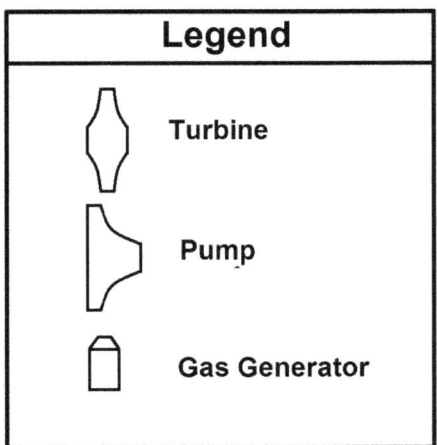

Another approach to the gas generator cycle is to use a monopropellant like hydrogen peroxide to drive the turbine.

Gas generators do generate a small percentage of thrust and sometimes these are used to settle the propellants after the rocket engine is shut off. This cycle is used for both booster stage vehicles as well as in-space vehicles.

Expander Cycle

The expander cycle uses the heat of the engine to vaporize a propellant and drive the turbine. The gases from the turbine are then fed back into the combustion chamber where they are burned. Because no propellant is thrown overboard, and all of it is burned in the combustion chamber, this approach is much more efficient than the gas generator cycle. Additionally, the gases driving the turbine are relatively cooler than with other rocket cycles.

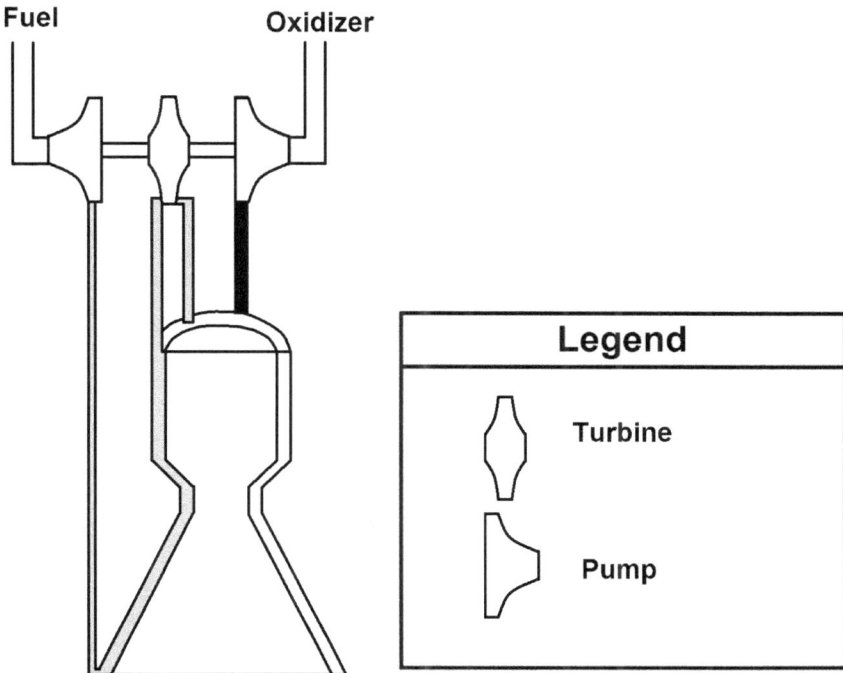

Fuel   **Oxidizer**

Legend

Turbine

Pump

The expander cycle can't work for all propellants. Some propellants don't vaporize cleanly or at the right temperatures. It is also limited in its maximum power by the heat available to vaporize the turbine driving propellant.

Staged Combustion Cycle

Staged combustion cycles burn some amount of propellant to drive the turbopump and then introduce the burned gases into the combustion chamber to increase engine performance. It's similar to a gas generator cycle but the turbine exhaust is fed back into the combustion chamber. Because a large amount of propellant is available to be used to drive the turbine, it can have very high pumping pressures with relatively lower stress on the turbines. When an oxidizer like liquid oxygen is used to turn the turbine, these gases can be designed to have relatively benign temperatures. But, an oxidizer-rich approach generally requires special coatings to avoid corrosive and ignition effects at higher temperatures.

Fuel   **Oxidizer**

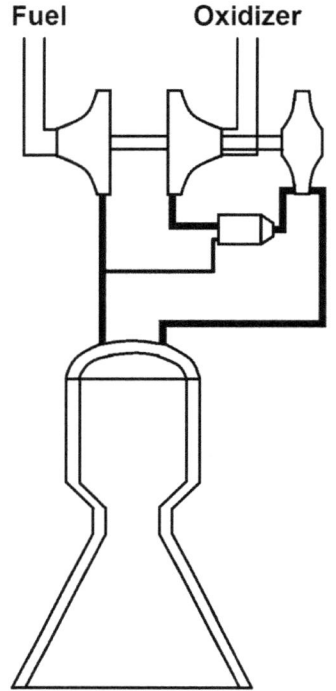

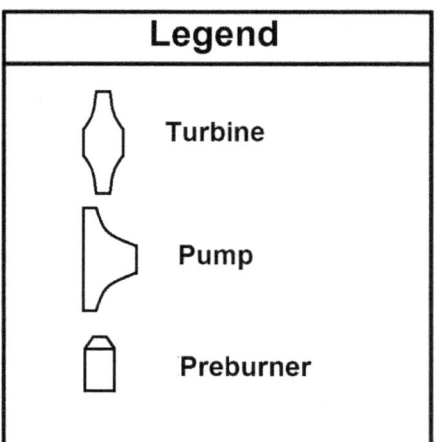

Examining the diagram of this cycle, we see that all of the oxidizer and a small amount of fuel are burned in the preburner to produce the gases that drive the turbine. The resulting gases from the turbine are then fed into the combustion chamber. This increases performance in two ways. First, all of the propellants go through the combustion chamber. Second, the oxidizer is heated by burning with the fuel in the preburner and this higher temperature oxidizer burns more effectively. Variations on this theme preburn both the oxidizer and fuel in separate preburners.

Microlauncher Missions and Vehicles

Chapter 8 – Microlauncher Missions

Now that we understand some of the technical basis behind the microlauncher idea, let's look at some of the things that we can do with microlaunchers and microspacecraft. In this chapter, we'll look at a wide array of possible missions that could be done with small spacecraft launched from small launchers.

We'll look at the types of missions that can be done based on the performance of the launcher and then adding considerations of added performance from the spacecraft. First we'll look at what can be done with microlaunchers with less than escape capability, then just over escape capability, and finally enhanced missions with full escape and trajectory control.

Under-Escape Missions

There are many useful missions that can be considered with microlaunchers that have under escape capability. For reference, this would likely be microlaunchers with a total dV capability of less than about 11.2 km/s.

High Altitude Missions

With microlaunchers capable of less than about 4.6 km/s dV, we can expect boosters that are able to reach the edge of space. This is the standard low-altitude sounding rocket regime and small microlaunchers can fulfill these types of missions just as well as existing sounding rockets.

What can be done with this kind of capability? One can demonstrate the basic ability to lift a payload to the edge of space and one can carry useful payloads to the edge of the atmosphere. As shown by our demonstration of the Super Arcas and Astrobee D sounding rockets, launchers of this size have real scientific and exploration capability.

Sounding rockets of this size can raise small payloads to the edge of the atmosphere able to do photographic, astronomic, and micro-gravitational research as well as upper atmosphere research. The payloads for these microlaunchers will be approximately what the upper stages of a multistage microlauncher would weigh.

A microlauncher first stage alone should be able to fulfill many of the requirements for a sounding rocket able to do these experiments. One major benefit of the liquid propellant approach is that it is easier to get lower acceleration and vibration than is usually seen with solid rockets. With even one upper stage, the payloads will be launched to very high altitudes and be suitable for longer-duration missions.

Earth Orbital Missions

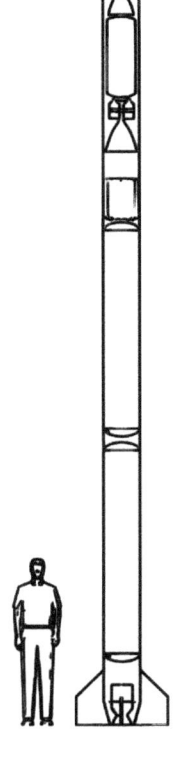

For microlaunchers with a total dV across all stages approaching just less than 10.6 km/s, there is the potential to reach orbit around Earth. This opens up a whole slew of mission types. There are Earth observation missions, communication missions, astronomical missions, and microgravity missions.

Orbital missions are demanding because they require highly accurate guidance systems. Gone are the days when one could launch satellites into arbitrary rough orbits. With many expensive assets in space, it is likely one would have to prove an ability to place the payload into a particular orbit with tight tolerances. Nonetheless, microlauncher sized orbital launch vehicles are feasible.

What might an orbital capable microlauncher look like? Let's look at a very conservative design example for a 1U cubesat to a very low altitude orbit of 200 km. Because of this low altitude, this satellite would only be expected to stay in orbit for at most a few days.

The size of the vehicle shown here probably represents one of the largest scale vehicles which can properly be called a microlauncher. Building and operating this vehicle is a large effort. But it is worth considering what the upper scale of microlaunchers vehicles might look like.

The following table lists the specifications of all of the stages.

Parameter	Stage 3	Stage 2	Stage 1	Units
Payload	1	5.447	61.305	Kg
Propellants	Lox/Propane	Lox/Propane	Lox/Propane	
Average I_{sp}	325	335	249	S
Stage dV	2789	4679	2743	m/s
Payload Ratio	0.225	0.098	0.110	
Structural Coef	0.286	0.167	0.251	
Propellant Ratio	0.714	0.833	0.749	
Mf/Me Ratio	2.399	4.154	3.075	
Propellant Mass	3.177	46.549	417.796	kg
Stage Empty Mass	1.271	9.310	140.011	kg
Me	2.271	14.757	201.317	kg
Mf	5.447	61.305	619.113	kg
Thrust	223	1446	12135	N

Each stage will be described in some detail in the following paragraphs.

Stage 3

The third stage is a relatively small vehicle, about 1.27 kg empty, 5.4 kg fully loaded. In order to accommodate a guidance system, the stage has a relatively low propellant ratio and therefore it has a relatively small delta V for a stage operated in a vacuum. With a vehicle this small, things like guidance are difficult to make much smaller.

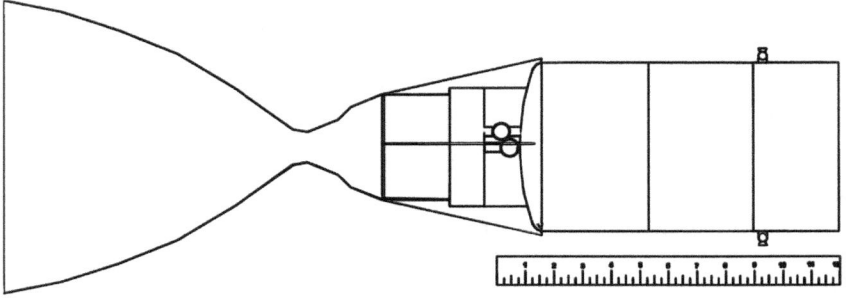

This stage uses liquid nitrogen in a container in the liquid oxygen tank to provide pressurization for both propellants and attitude control. In this situation, the nitrogen's vapor pressure is exactly right for propellant tank pressurization. The guidance and control system keeps the stage oriented in its flight. Welded aluminum tanks hold chilled propane and liquid oxygen as propellants. Small, light pyrotechnic valves are used to initiate combustion. The low-pressure engine is regeneratively cooled with a large 100:1 expansion ratio nozzle to provide high vacuum efficiency. The nozzle would likely be made from sheet steel using metal spinning. This stage provides attitude control for both itself and for stage 2 and uses the same liquefied nitrogen pressurant as is used to pressurize the tanks for cold gas.

Stage 2

Stage 2 looks similar to stage 3 but it has no guidance system. It has one ablatively cooled rocket engine with a thrust of about 1446 N. In order to get high performance that fits within the first stage diameter, the engine has a P_c (chamber pressure) of about 896 kPa. It also has a large 100:1 nozzle to give high vacuum performance.

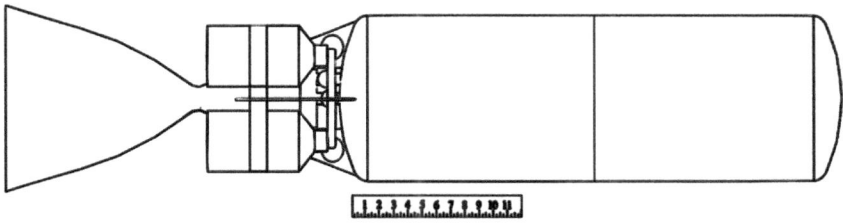

It uses off-the-shelf paintball pressure tanks to pressurize the propellants. It's tanks are also made from welded aluminum. Pyrotechnic valves and igniters are used to initiate combustion.

Stage 1

Stage 1 is the launch vehicle for the upper stages. It has a diameter of 39 cm and a total height of 10 meters for a length to diameter ratio of about 25:1. It also uses chilled propane and liquid oxygen for propellants. Most of the structure is made of welded aluminum. Paintball tanks with helium provide the pressure for the propellants.

It uses a single gimbaled engine with a thrust of about 12135 N and a chamber pressure of 1035 kPa. The engine has a nozzle expansion ratio of 3.3 providing about 214 seconds of I_{sp} at sea level and 274 seconds in a vacuum, with an average I_{sp} in flight of 249 seconds.

The recommended flight trajectory for this vehicle is a vertical flight with ejection of the upper stages above 100 km altitude. This ensures that the first stage avoids passing over populated areas.

Summary

This high-level conceptual design shows what can be done with microlaunchers under 1000 kg total liftoff mass. The numbers used in this conceptual design are realistic and there is sufficient margin for each of the stages' masses that it is likely that a stage like this could be built.

For smaller payloads, the overall liftoff mass can be significantly less than this example and the rocket size can be much smaller. In this example, each stage had quite high empty mass (and thus low propellant ratios). Dedicated development efforts should allow the empty mass to become substantially lower and therefore, the vehicles overall size will be quite small. Fifteen centimeter and twenty centimeter diameter orbital vehicles might be possible for very small payloads.

But LEO is not Exploration Any More

But Low Earth Orbit (LEO) has been done hundreds of times and it's hardly exploration any more. Therefore, in the next few sections, we'll examine possible missions beyond LEO. However, to do that, the overall rockets will need to have escape velocities. Robert A. Heinlein is credited as saying that "Once you get to earth orbit, you're halfway to anywhere in the solar system." The microlauncher variation of this is that "...once you get to Earth orbit, you're ONLY halfway to anywhere in the solar system." We need to push ourselves to go the rest of the way.

Just Over Escape Velocity Missions

By lowering the payloads on the stages, it is possible to increase the overall vehicle velocity attainable into the escape velocity range: just over 11.2 km/s. It is also possible to add an extra stage to increase the overall attainable velocity. Using the orbital launch vehicle as an example, it could lift a payload of 61 kg to 200 km altitude. Let's look at what size the stages would be (and the resulting payload) using the orbital first stage launcher with 3 upper stages going about 100 m/s over escape velocity.

The upper stages must be able to deliver about 11.27 km/s. Presuming the 3 upper stages all operate in a vacuum, and dividing the required dV equally, each stage would have 1/3 of 11.27 km/s or about 3.76 km/s for their stage dV's. If we presume that each stage has an I_{sp} of 325 seconds, and that they each have propellant ratios of 80%, then we can calculate the stage dimensions. The following table shows each stage's characteristics.

Parameter	Stage 4	Stage 3	Stage 2	Units
Payload	0.150	1.750	10.365	kg
Propellant Isp	325	325	325	s
dV	3760	3760	3760	m/s
Payload Ratio	0.155	0.155	0.155	
Structural Coef	0.200	0.200	0.200	
Propellant Ratio	0.800	0.800	0.800	
Mf/Me Ratio	3.253	3.253	3.253	
Propellant Mass	0.774	5.769	42.985	kg
Stage Empty Mass	0.194	1.442	10.746	kg
Me	0.344	2.560	19.075	kg
Mf	1.118	8.329	62.060	kg

Therefore, stages with the above characteristics would be able to put a payload of 150 grams onto an escape trajectory. It's a small payload, but with the advanced capability of electronics, we can put substantial functionality into that payload. That's basically the size of a smartphone. What could you do if you had a smartphone on an escape trajectory? We'll discuss some ideas in an upcoming chapter.

Simple Escape Trajectories

Before we look at ideas for missions, we should understand how to get a payload onto an escape trajectory. In an earlier chapter, we discussed the mathematics of the escape trajectory and stated that it was 1.414 times an orbital velocity. For the most part, to go on an escape trajectory, you should ascend as if you're going into a particular orbit, but rather than enter that orbit, at the point you would normally be in orbit, make sure that you're going 1.414 times that velocity. That'll put you on an escape trajectory away from Earth.

The following diagram shows one approach to going into an escape trajectory for a three stage vehicle but is meaningful for any number of stages.

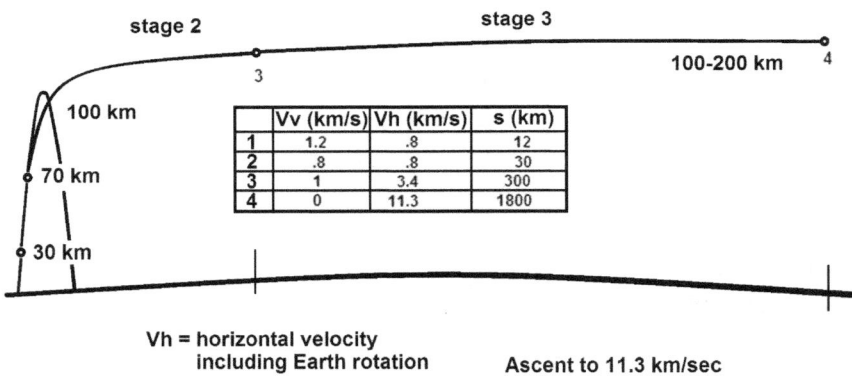

	Vv (km/s)	Vh (km/s)	s (km)
1	1.2	.8	12
2	.8	.8	30
3	1	3.4	300
4	0	11.3	1800

Vh = horizontal velocity
including Earth rotation   Ascent to 11.3 km/sec

In this trajectory, the rocket lifts off from Earth and at 30 km, the vehicle has a vertical trajectory of 1.2 km/s and a horizontal Eastward velocity of 0.8 km/s at burnout of the first stage (point 1). It continues coasting until point 2 is reached and it releases the upper stages. The second stage ignites at point 2 and imparts sufficient vertical velocity to raise the payload up to 200 km. The first stage follows a ballistic trajectory to hit about 120 km downrange. After this vertical velocity is added to stage 2, all remaining velocity is directed tangentially to the Earth's surface (horizontally). At point 3, the 2nd stage burns out and the third stage ignites putting all of its velocity horizontally (tangentially). Finally, at point 4, with an altitude of 200 km, the third stage burns out and leaves the payload with 11.3 km/s tangential velocity. This is sufficient to put the payload on an escape trajectory.

As it leaves Earth, the upper stage and spacecraft will be affected by the Earth's and the Sun's gravity and could enter into a solar orbit as shown below. This shows a view looking down on the North pole of the Earth in its trajectory around the Sun.

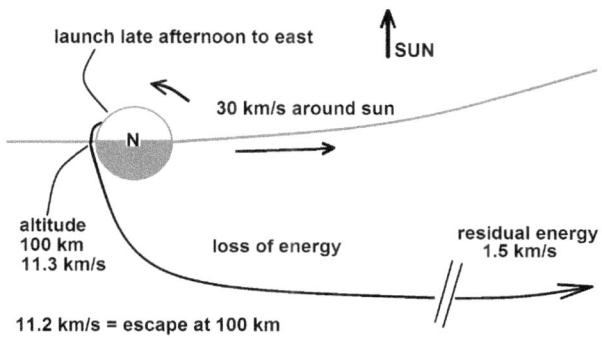

A spacecraft launched on an escape trajectory in the late afternoon as shown above would begin losing energy as it traveled away from Earth and would assume a solar orbit some distance from Earth. It would have a velocity of about 1.5 km/s relative to Earth. Over month's duration, it could be expected to continue on a solar orbit as shown below.

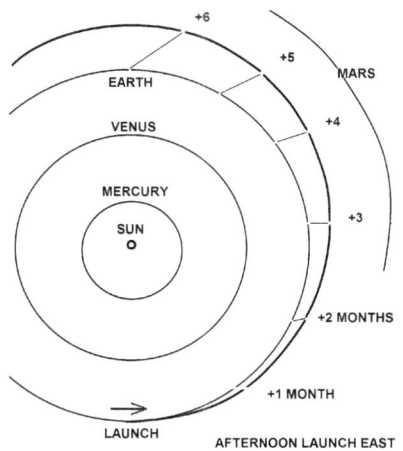

For months after the launch, it would be visible in the night sky for durations of hours. This would allow long-term opportunity to establish and maintain communications with the spacecraft.

Escape Without Accuracy

Given a microlauncher with the ability to enter to an escape trajectory but without any significant accuracy, there are a number of useful things that can be done.

Test and Demonstration Flights

The first flights with vehicles able to reach escape velocity without any significant accuracy or in-flight control will likely be demonstration as test flights. These kinds of flights will allow the developers to validate their systems, establish performance and reliability. In short, they will demonstrate the feasibility and capability.

These rough flights will also allow demonstration of monitoring and communication equipment. It is likely that a laser communication link can be demonstrated allowing high bandwidth and reliable communications with ground stations. One of the beauties of escape trajectories is that the vehicle will remain visible in the sky for hours at a time each day or night. With highly sensitive communications systems, bidirectional command and data download is possible.

Revenue Flights

However, these first rough capability flights also begin to lend themselves to revenue generating flights. Space videos and pictures can be generated; there is always a desire and need for good pictures of Earth. However, it might also be possible to do "vanity" revenue flights by sending items like personal items and business cards on escape trajectories. Space burials of ashes might also be possible with these kinds of vehicles.

Solar Particle Observation Mission

Real science work is also possible with even very small spacecraft. If simple particle density sensors are placed on tiny spacecraft, then densities of solar wind can be measured. Since these kinds of vehicles are relatively cheap, they might be produced by the dozens and they could give continuous and widely separated samples of solar activities around Earth.

Extra-Solar Planet Survey

With some good optics and accurate orientation control, it is possible to have spacecraft which can stare at single stars for long durations. This can be useful for extra solar planet surveys. Using silicon mirrors of about 10 cm diameter and cool, temperature stabilized photodiode optical sensors, these spacecraft can stare continuously at one star and plot starlight to a high precision from which it would be possible to plot minor variations over time. Using small laser communications links, this data could be relayed back towards Earth or to a LEO satellite.

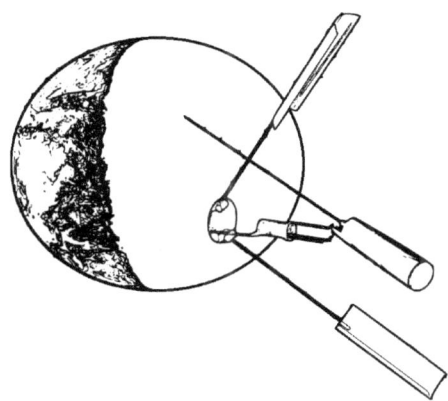

The above image shows how a small space telescope can relay images back to Earth using laser communications. The telescope imager can be used to maintain stability on the Earth which will appear as a thin crescent throughout most of such missions. Small paddles can use solar wind to facilitate spacecraft orientation and stability without propellant.

Rogue Planet Search

It may also be able to perform rogue planet searches using gravity lensing to detect bodies passing in front of stars. By using small compact star clusters, it may be possible to detect more than one event as the object passes two or more stars. A slightly defocused image on a grey-scale high resolution Charge-Coupled Device (CCD) imager can spread starlight over several pixels. This can be used to measure and record brightening and dips in star light indicative of an object passing in front of a star. By correlating related events from several stars, it may be possible to detect planets traveling in interstellar space not in orbit around a star.

High Lunar Orbit

With a basic escape trajectory, it may also be possible to place a spacecraft in a high lunar orbit. With just slightly over escape velocity, this would likely be a highly elliptical orbit, but it is possible. Once done, these spacecraft can do lunar imaging and other studies like solar wind flux measurements.

Asteroid Fly-By Missions

Although this kind of mission will require some basic propulsion to refine the trajectory and fly-by distance, it is possible to send small spacecraft on pass-by missions of asteroids, imaging them and helping to identify the asteroids' orbits more accurately.

Escape Missions with Course Correction and Propulsion

If the small spacecraft have some basic course correction and propulsion capability, then even more sophisticated missions are possible.

More Sophisticated Asteroid Fly-By's

With some basic course correction and propulsion, far more sophisticated asteroid fly-by missions are possible. With this capability a spacecraft can rendezvous and orbit an asteroid and take images and video. It may also be possible to "tag" an asteroid with a beacon to facilitate tracking from Earth.

Accurate Lunar Orbits

With a tiny bipropellant rocket, it is possible to place a spacecraft in fairly accurate lunar orbits. Low lunar orbits are possible where detailed images of lunar surface features can be taken.

Electric Propulsion Missions

Electric propulsion systems on tiny spacecraft open up tremendous capabilities. Using solar cells to power them, significant velocity changes and trajectory changes are possible. Ion propulsion systems can provide upwards of 5000 seconds of I_{sp} in very tiny packages. This will allow all of the earlier described missions to be performed with great accuracy and flexibility.

Solar Sail

Solar sails use solar light radiation to provide propulsion. Since they use no propellant, they can enable extremely long duration missions under sophisticated control. In cases where tiny spacecraft are not be able to carry stiff structural members for holding the sails, the spacecraft can be rotated to provide structural stiffness to the sails. This capability has already been used on several cubesats in the past. Pressurized gases may also be used to provide sail structure and stiffness. The sail material itself need only be extremely thin aluminized plastic materials.

Enhanced Missions

Even though microlauncher spacecraft are tiny, their potential is boundless. Within theoretical possibilities are missions involving landing on other planetary bodies such as the moon and asteroids. They may even be utilized in part of larger activities such as asteroid deflection. In this case, they might be the carriers of materials to paint the asteroid with reflective or absorbent materials resulting in sufficient orbital deflection to miss Earth. In other cases, they might be used to plant flags on asteroids or other planetary bodies. Used for tagging asteroids, they can become an important tool for Earth's defense against an asteroid collision.

Chapter 9 – Upper Stage Development

Upper stages of microlaunchers require special attention to be able to meet the requirements for the kinds of performance needed. They are going to be physically small, they must be lightweight and some parts will require specialized construction techniques. But because upper stage vehicles operate in or very near a vacuum, they don't have to face the same structural and drag effects caused by the atmosphere like booster stages. Additionally, since the vehicle is flying through a near-vacuum, the rocket engines are able to operate much more efficiently because they can take advantage of larger nozzles.

Upper Stages, Air Pressure, Air Density and Dynamic Pressure

In chapter 5, an approximation for air pressure versus altitude was introduced:

$$p = 101.353 \cdot e^{-0.000138714 \cdot a}$$

Where **p** is the pressure in kiloPascals and **a** is the altitude in meters.

Another important aspect of the atmosphere is its density. An approximation of the density of the atmosphere versus altitude is:

$$d = 1.42127 \cdot e^{-0.0001411 \cdot a}$$

Where **d** is the density in kg/m^3 and **a** is the altitude in meters.

Again, neither of these equations is meant to be true, accurate representations of the atmosphere's properties, but they are good enough approximations and definitely represent the trends well enough. See the US Standard Atmosphere 1976 table included in Appendix B for more accurate values. But it is convenient to have equations with which to work.

Upper stages might separate from the first stage at altitudes as low as

60 km but are more likely to stage near 100 km. Working with these equations, we can begin to understand some of the implications of altitude effects for the flight of upper stages. Obviously, once above about 100 km, the spacecraft is considered to be "in space," but between the time that the first stage burns out and it reaches this altitude, it is worth understanding the nature of the aerodynamic effects on the vehicle. The following table shows the pressure and density at various altitudes likely to be encountered by upper stages based on the 1976 Standard Atmosphere.

Altitude (km)	Pressure (kPa)	Density (kg/m^3)
60	0.02196	0.00030968
70	0.00522	0.00008283
80	0.00105	0.00001846
90	0.00018	0.00000342
100	0.00003	0.00000056

As can be seen in this table, both the pressure and density are quickly approaching zero (although they never reach there) as the altitude rises above 60 km. But, the pressure at 60 km is still 732 times that at 100 km and the density at 60 km is 553 times that at 100 km.

At 60 km, because of the speed that the upper stages are likely to be traveling, there can still be significant drag forces. As the vehicle travels through the atmosphere, it experiences an effect caused by drag known as **dynamic pressure**. Dynamic pressure is caused by the air in front of the vehicle being compressed; this compression spread over the frontal area causes a force which is measured just like air pressure. The equation for calculating dynamic pressure is:

$$q = \frac{1}{2} \cdot \rho \cdot v^2$$

where **q** is the dynamic pressure in Pascals, **p** is the air density in kg/m^3 and **v** is the velocity in m/s. Although the air density at 60 km is small, the velocity of the vehicle at this altitude is likely to be great and, therefore, the dynamic pressure is not insignificant. Let's look at an example. Suppose that the upper stage is at 60 km and is moving upwards at a velocity of 1650 m/s. Using the density from the table above, we see that the atmospheric density is $0.000310 \, kg/m^3$. Using the dynamic pressure equation, we determine that the dynamic pressure is:

$$q = \frac{1}{2} \cdot \rho \cdot v^2$$

$$q = \frac{1}{2} \cdot 0.000310 \cdot 1650^2$$

$$q = 421.9 Pa$$

This is not an insignificant pressure on the front side of the ascending upper stages; if the frontal area of the spacecraft is about 183 cm² (a diameter of 15.25 cm) then there is a force of 7.7 Newtons on the front of the vehicle at 60km. However, by the time the vehicle has reached 100 km, at the same speed, the dynamic pressure only produces a force of 0.014 Newtons.

The main point in introducing dynamic pressure is that even though the pressure and density conditions at staging altitudes is low, the aerodynamic forces experienced by the spacecraft can still be substantial if the velocity is high. Therefore, the microlauncher mission designer must be considerate of the altitude and trajectory of the rising upper stages while still in the upper limits of the atmosphere.

Optimization of Pressure Fed Stages with Pressure

It has been established that the air pressure at staging altitudes is low, and therefore we can use a large rocket engine nozzle to get higher performance without facing the problem of flow separation. Most likely, upper stages will utilize the pressure fed approach because higher chamber pressure produces less benefit than larger nozzles in a vacuum. But, given the choice of a pressure fed upper stage, at what pressure should a microlauncher designer make the upper stage?

With what has been reviewed so far, we can answer that. The major components that drive upper stage mass and performance are the pressurization system tank, the propellant tanks and the rocket engine. We can make simplifying assumptions about the construction techniques of these components which allow us to get a good approximation of their masses for given pressures.

Rocket Engine   **Oxidizer Tank**   **Fuel Tank**   **Pressurization Tank**

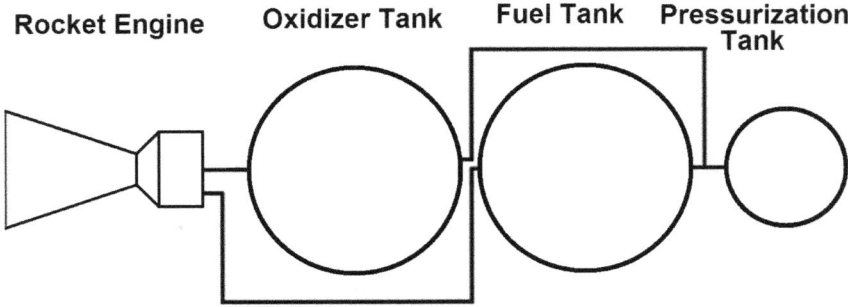

In chapter 6, the equations for propellant tank stress and mass estimation were introduced. Those can be used to determine the mass of propellant tanks for a given stress and pressure. If we fix the amount of stress that the tank faces while varying the pressure, we can derive an equation that represents the mass of the propellant tanks for a given pressure. If we simplify the tanks' designs to be spherical and of equal size, made of 6061-T6 aluminum with a safety factor of 1.5 for a fixed volume, then we can create a graph which plots the mass versus the pressure. The specific values used reflect the 2nd stage of the 'Just Over Escape Velocity Mission' system described in Chapter 8.

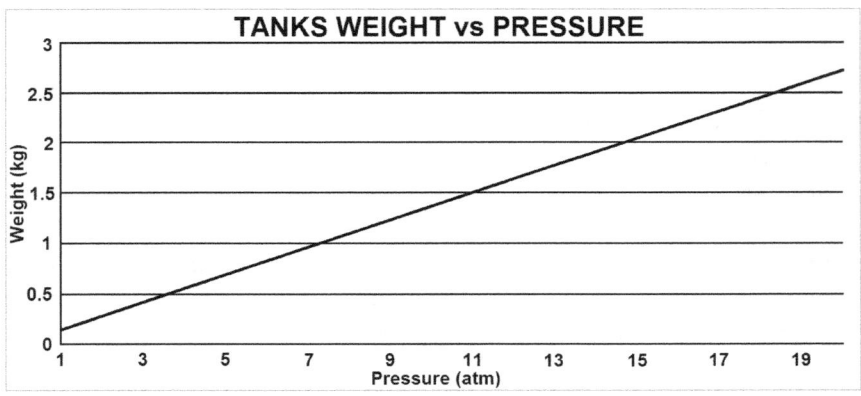

As this graph shows, the higher the pressure, the heavier the tanks are. That's not too difficult to imagine. The tanks have the same internal volume, are basically the same size and shape and they only need to be thicker to hold the increased pressure for the same stress.

In much the same way, we can calculate the dimensions and then determine the mass of the rocket engine that would operate at some percentage of the propellant tank pressure for a fixed I_{sp} in vacuum. It is a bit more complicated to compute because we're working backwards from a desired engine thrust and I_{sp} performance, determining the size of the combustion chamber and nozzle that would produce that performance (with a nozzle size which varies with chamber pressure) and then estimating the mass, but it can be done. The general trend for the engine is that as the combustion chamber pressure gets higher, the engine gets smaller and the nozzle has a lower expansion ratio; but, the higher pressure requires thicker walls to contain. The following graph shows the trend.

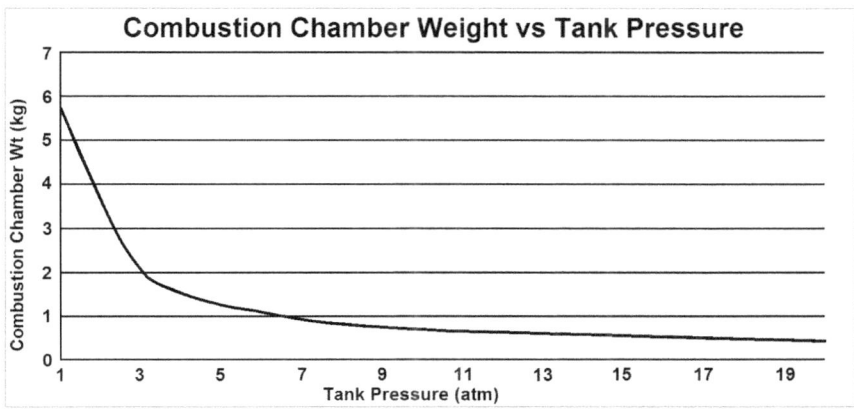

The basic trend observed is that the engine gets lighter with increased pressure, although not completely linearly. The main reason for this is that the nozzle is larger at lower pressures.

We can also estimate the mass of the pressurant system for a stage. Assuming a pressure tank performance factor of 6350 meters and 20 MPa helium pressurant, the following graph shows the mass of the pressurant system against the propellant tank pressure. Obviously, higher pressure requires heavier pressurant tanks to hold it and more pressurant is required.

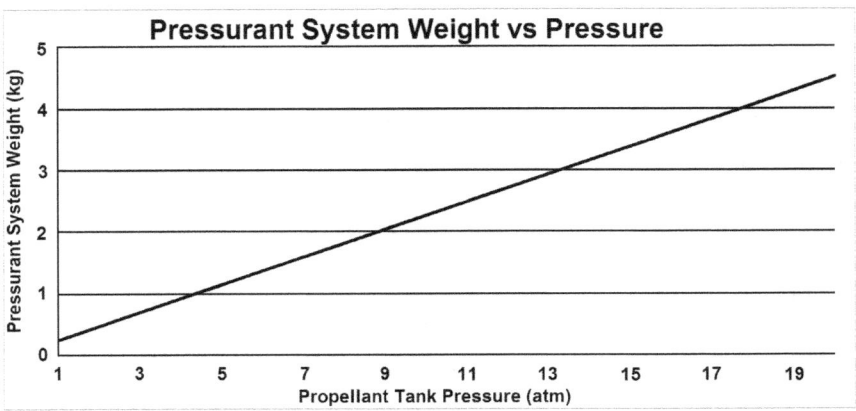

Since a pressure fed tanks' masses greatly overshadow the combustion chamber's mass, the resulting sum of all of these components primarily increases with pressure. The following graph shows how these trends work together. The stage mass is the sum of the pressurant system tank, the propellant tanks and the rocket engine plotted against the propellant tanks' pressure.

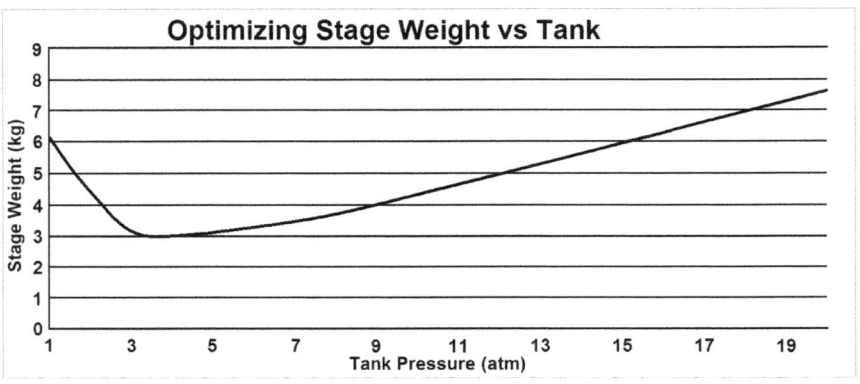

It shows that mass is minimized near 3 or 4 atmospheres; this range should be selected for a rocket stage that operates entirely in a vacuum to get the lowest stage mass.

Construction Techniques for Upper Stages

Upper stages of microlaunchers require careful consideration to meet the mass and performance goals. Reviewing the parameters for the escape launcher upper stage, we see the following numbers.

	Stage 4	Stage 3	Stage 2	Units
I_{sp}	325	325	325	seconds
Payload Mass	0.150	1.617	9.981	kg
Propellant Mass	1.120	6.913	42.667	kg
Stage Empty Mass	0.347	1.452	8.960	kg

If we presume that we're using liquid oxygen and butane as the fuels with liquid nitrogen as a pressurant, here are the characteristics of those propellants at their operating conditions.

	Liquid Oxygen	Liquid Butane	Liquid Nitrogen	Units
Temperature	90	295	90	Kelvin
Density	1.141	0.604	0.746	g/cc
Vapor Pressure	100	202.65	350	kPa

Using these properties of the propellants, then we can estimate a more complete properties table. Additionally, knowing the propellant masses and densities, we can calculate the volumes of the tanks that hold them.

	Stage 4	Stage 3	Stage 2	Units
I_{sp}	325	325	325	S
Payload Mass	0.150	1.617	9.981	kg
Propellant Mass	1.120	6.913	42.667	kg
Stage Empty Mass	0.347	1.452	8.960	kg
OF Mixture Ratio	2.1	2.1	2.1	
Oxidizer Mass	0.759	4.683	28.903	kg
Fuel Mass	0.361	2.230	13.763	kg
Oxidizer Volume	665	4104	25332	cc
Fuel Volume	601	3712	22913	cc

Now that the tanks volumes are known, their individual diameters can be determined. One should add a little bit more spare volume (known as "ullage") to each tank to allow for expansion of the fluids. It is prudent to allow about 5% extra. Therefore, we derive the following tank dimensions.

	Stage 4	Stage 3	Stage 2	Units
Ox Volume w/ullage	698	4309.2	26599	cc
Fuel Volume w/ullage	631	3897.6	24059	cc
Ox Sphere Diam	10.98	20.18	37.02	cm
Fu Sphere Diam	10.64	19.52	35.80	cm

Knowing the diameters of the propellant tanks, we can determine the tensile stress that each is under as a function of pressure. If we select a tank material first, say steel with a tensile strength of 415 megaPascals and use a safety factor of 1.5, then we can determine the thickness of the material for the tanks. Reviewing the spherical stress equation from chapter 6:

$$\sigma = \frac{p \cdot r}{2 \cdot t}$$

$$t = \frac{p \cdot r}{2 \cdot \sigma}$$

For mathematical and construction simplicity, I will round the tank diameters upwards a bit and make the oxidizer and fuel tank diameters the same.

	Diameter (cm)	Pressure (kPa)	Stress (MPa)	Tank Skin Thickness (mm)
Stg 2 Oxidizer Tank	40	345	275	0.125
Stg 2 Fuel Tank	40	345	275	0.125
Stg 3 Oxidizer Tank	21	345	275	0.066
Stg 3 Fuel Tank	21	345	275	0.066
Stg 4 Oxidizer Tank	11	345	275	0.035
Stg 4 Fuel Tank	11	345	275	0.035

As can be seen, at these low pressures, the propellant tank thicknesses are very thin. Is it even possible to construct tanks with these kinds of tiny thicknesses?

There are limits to standard construction techniques for the tanks. The usual way that propellant tanks suitable for materials like liquid oxygen are built is using what is known as ***metal spinning***. It involves spinning a plate of metal over a positive shape, known as the mandrel, and pressing or forcing it to take the shape of the mandrel. Once two halves are made, they are often welded together. But there are limits to how thin the material can be made using metal spinning and welding. For most microlauncher upper stage tanks, their thicknesses are probably too thin for metal spinning.

It might be possible to just increase the thickness of the tank material, making the tanks far thicker than they need to be, but even that requires increasing tank thickness upwards of four times. Therefore, the tanks are four times as heavy as they need to be. In the smallest tanks, stage 4's oxidizer and fuel tanks, it would be necessary to increase the thickness and mass up to 20 times before they become reasonable for being made with metal spinning. We might also try a different material that can take less stress like aluminum and thereby make the tanks thicker. However, even that is too thin for metal spinning.

There is an alternative approach that is advocated for microlaunchers: electroforming. Electroforming is essentially identical to electroplating but with thicker plating layers. Electroforming, like electroplating, involves using an electric current to deposit a metallic layer onto a conductive form. Electroplating is likely to be an enabling technology for microlaunchers because complex, light, functional parts can be produced relatively easily.

Electroplating Propellant Tanks

Nickel is a widely used and available electroforming and electroplating material. It can be deposited over conductive surfaces in very thin layers and yet produces bodies with high tensile strength. It can be expected that nickel electroformed tanks will easily demonstrate yield strengths above 415 MPa. The highest yield strengths seen for electroformed nickel are above 1300 MPa.

The development process for nickel electroformed tanks involves forming a mandrel that is (or is made to be) conductive, placing it in the electroforming chemical bath, applying the appropriate voltages and currents for the proper duration of time, removing the plated mandrel from the bath, cleaning and removing the mandrel.

Electroplating Rocket Engines

It is also possible to electroform rocket engines and engine nozzles for small microlauncher rocket engines. In fact, this is one of the more common techniques used on many commercial launchers; a combustion chamber and nozzle are often produced using a machined liner with cooling channels, and then a closing jacket is usually electroformed over the liner. Therefore, although the idea of electroforming rockets might seem novel, it is in fact an industry standard method for producing rocket engines.

Electroplating Valves

Finally, since microlauncher upper stages will require valve mechanisms to control propellants, this is something that electroforming can help with as well. Bellows are often used to seal pressure pistons in some actuators and they are very often made by electroforming in commercial practice. With careful design, it is possible to produce small lightweight, pressure activated valve mechanisms for microlauncher upper stages as well.

An Electroforming Process

A quick review of how one might electroform propellant tanks is looked at here.

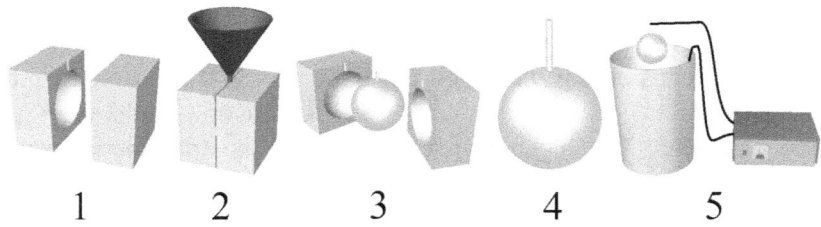

In the first step, a mold is made from a material like plaster or resin. Next, the mold halves are closed and sealed and then the mandrel material is poured into the mold. Suitable mandrel materials are wax, resin or Wood's Metal. The mandrel material can be rotationally molded in the mold cavity and thereby create a thin-shelled mandrel. Next, after the mandrel is cured, it is removed from the mold. In the next step, the mandrel is prepared for plating. Non-metallic mandrel materials like wax or resin can be coated with conductive material to make the mandrel electrically conductive. If the mandrel is made from a conductive material like Wood's Metal, then it is already conductive. The mandrel is placed on a suitable holding shaft that will allow it to be dipped in the electroforming solution. Finally, the mandrel is placed in the electroplating solution and an appropriate electric current and voltage is applied to it. After a suitable electroforming time, the mandrel can be removed from the plating solution. It is now ready for finishing for use as a propellant tank. Wax or Wood's Metal can be melted out of the newly formed propellant tank.

Using electroforming, it is possible to very carefully control the thickness of the plating. By controlling the current and the amount of time in the plating solution, the thickness is established. Nickel electroforming can produce quite light tanks with high tensile strengths.

Beverage Containers as Upper Stage Propellant Tanks

Another approach to producing small propellant tanks for upper stages is to utilize aluminum beverage cans. This might seem like a far-fetched idea, but it has actually been demonstrated as feasible.

Commercially produced soda containers have very thin walls and are extremely lightweight, yet it is possible to use them as pressurized propellant tanks; they are often designed to support about 620 kPa of pressure as they arrive from the manufacturer. Either normal soda cans or larger energy drink cans may be used with little modification. They can easily be brazed together with end caps and propellant feed lines using simple hobbyist quality butane torches. They might also be glued together with epoxy or have their fittings glued with epoxy but this is less likely to survive liquid oxygen temperatures.

An Example Upper Stage

A small upper stage based on beverage cans will be shown. This represents a low-tech, relatively high mass approach, but it is one which is very informative to consider. For this design, a commercially available aluminum beverage container with a twist-on cap will be considered. The following sketch illustrates the basic design. The cans are about 21.5 cm tall, 7 cm in diameter and are manufactured to hold 710 mL of fluid. Based on measurements with fittings, each can with a pressure feed tube and a propellant withdrawal tube weighs about 34 grams. The walls are about 101 microns thick and the manufacturers specify that they can hold about 620 kPa of pressure.

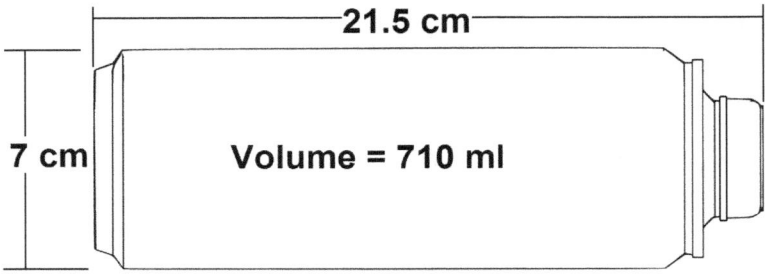

This design experiment will consider the fourth stage of the following system, which is suitable for the first stage booster described in earlier chapters. This stage can be used as the final stage of an Earth escape mission with a 125 gram spacecraft.

	Stage 4	Stage 3	Stage 2	Units
Oxidizer	LOX	LOX	LOX	
Fuel	Butane	Butane	Butane	
OF Ratio	2.1	2.1	2.1	
Isp	325	325	325	s
deltaV	3759.1	3759.1	3759.1	m/s
Payload Mass	0.125	1.617	9.981	kg
Propellant Mass	1.120	6.913	42.667	kg
Stage Empty Mass	0.347	1.452	8.960	kg
Oxidizer Mass	0.759	4.683	28.903	kg
Fuel Mass	0.361	2.230	13.763	kg
Oxidizer Volume	665	4104	25332	cc
Fuel Volume	601	3712	22913	cc

This stage will use a 49 Newton rocket engine with a combustion chamber pressure of 275 kPa and an expansion ratio of 125:1. Because of this low pressure, the engine will be very thin-skinned construction, similar to the propellant tanks. It is likely produced using electroforming as a regeneratively cooled engine. Here is a plumbing diagram of the major components.

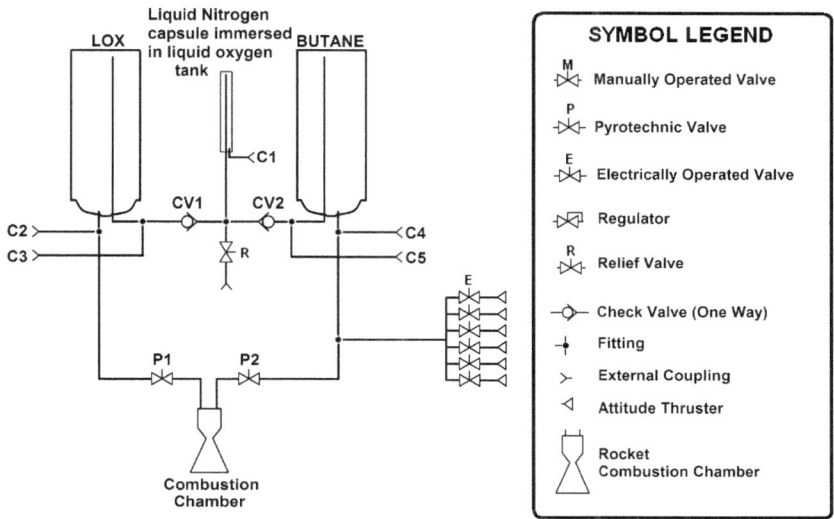

Pressurization is provided by a tube of liquid nitrogen immersed in the liquid oxygen. This will result in the liquid nitrogen having the proper vapor pressure of 3.5 bar to provide pressurization for both propellants. A bank of six small electrically operated valves with small nozzles on them provides attitude control using the butane propellant. Although this has a low I_{sp}, it provides sufficient functionality with low mass. A calculated amount of additional butane is kept in the fuel tank beyond what is needed for the main propulsion thrust. The following table gives a mass breakdown estimate and shows that the stage is in its mass budget with room to spare.

Qty	Item	Unit Wt (grams)	Extended Mass (grams)	Budget Percent
2	710 ml Cap Cans	34.0	68.0	19.6%
2	Liquid Nitrogen Pressurant @ 345 kPa	2.8	5.6	1.6%
1	49 Newton Thrust Rocket Engine	31.0	31.0	8.9%
1	Liquid Nitrogen Container	0.8	0.8	0.2%
6	microfluidic valves (attitude control)	4.5	26.9	7.8%
1	40 s Attitude Control Butane Propellant	26.7	26.7	7.7%
1	GNC Computer	60.0	60.0	17.3%
2	Button Battery 3.6 V	12.4	24.8	7.1%
2	check valves	5.9	11.8	3.4%
1	pressure relief valve	5.9	5.9	1.7%
6	Attitude Control Nozzles	0.5	2.7	0.8%
2	Propellant Valves	9.1	18.1	5.2%
100	cm of 3.8mm aluminum tube	0.2	22.0	6.3%
1	Thrust Structure	35.0	35.0	10.1%
		TOTAL	339.5	97.8%
		BUDGET	347.0	
		REMAINING	7.5	2.2%

Here is an image of what the stage might look like.

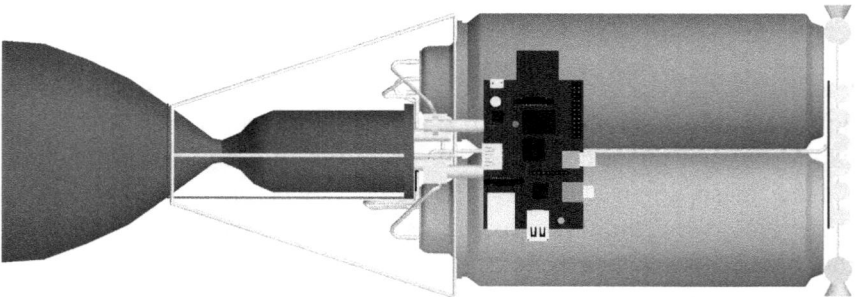

This design experiment demonstrates an initial feasibility for a small stage based on beverage cans with the necessary impulse and mass requirements at the sizes needed for microlauncher escape missions.

An Advanced Electroformed Upper Stage

We've seen that at low chamber and propellant tank pressures, the propellant tanks need only be 35 microns thick or less for upper stage tanks. However, most construction technologies often used for making pressure vessels are unable to make tanks that are that thin. In the last section, we looked at using aluminum beverage cans but even they are about 3 or 4 times as thick as are needed (and therefore 3 or 4 times heavier than necessary). Using electroforming, however, we can produce very thin, strong tanks. In this section, an upper stage with electroformed tanks will be examined. This stage represents an extremely advanced design and has amazing performance for such a small, light pressure fed stage.

The stage being examined will be the third stage of a 3 stage vehicle with escape capability. Because the third stage has such high performance, less performance is allocated to the second stage, whose job is merely to raise the 3rd stage above the atmosphere and add a little extra dV. The following table highlights the overall multistage vehicle's parameters in order to give a context for understanding the role of this stage. This section will not go into any more detail about stages 1 or 2; only stage 3 will be described.

	Stage 3	Stage 2	Stage 1	
Oxidizer	Lox	Lox	Lox	
Fuel	Butane	Propane	Propane	
Stage Extra	0.000	0.000	0.000	kg
Payload	0.200	12.824	32.693	kg
OF Ratio	2.200	2.200	2.200	
Oxidizer Density	1.064	0.970	1.141	g/cc
Fuel Density	0.662	0.662	0.538	g/cc
Avg Density	0.847	0.847	0.847	g/cc
Average Isp	320	320	249	seconds
Desired DeltaV	8614	2438	2743	m/s
Body:Fuel Ratio	0.05200	0.12500	0.33512	
Payload Ratio	0.016	0.645	0.110	
Structural Coef	0.049	0.111	0.251	
Propellant Ratio	0.951	0.889	0.749	
Mf/Me Ratio	15.563	2.175	3.075	
Propellant Mass	12.000	17.662	222.799	kg
Oxidizer Mass	8.250	12.142	153.174	kg
Fuel Mass	3.750	5.519	69.625	kg
Oxidizer Volume	7754	12519	134246	cc
Fuel Volume	5664	8337	129361	cc
Stage Weight	12.624	19.869	297.463	kg
MT	0.624	2.208	74.664	kg
Me	0.824	15.031	107.357	kg
Mf	12.824	32.693	330.156	kg
Stage Impulse	37631	55387	543674	N-sec
Cum delta V	8614	11052	13795	m/s

The first thing to notice is that the third stage has a large dV increment of 8.6 km/s; the overall vehicle's total dV is about 13.8 km/s.

The following diagram illustrates the basic layout of stage 3.

The mass estimate table lists all of the major stage components and their estimated masses. It shows that the design can meet its design mass goals while meeting overall system performance and still having budget margin to accommodate oversight.

Qty	Description	Unit Weight (grams)	Extended Weight (grams)
1	Lox Tank 3 Spheres 18cm Diam	121.4	121.4
1	Butane Tank 2 Spheres 18cm Diam	86.5	86.5
1	118 Newton Rocket Engine	115.7	115.7
2	Servos for Engine Valves	5.0	10.0
3	Servos for Attitude Control	5.0	15.0
2	Engine Propellant Valves	9.1	18.2
1	Butane Tank Filling Coupling	2.0	2.0
1	Liquid Nitrogen (30cc)	20.0	20.0
1	Liquid Nitrogen Capsule	2.0	2.0
1	Nitrogen Fill Coupling	2.0	2.0
1	Lox Tank Filling Coupling	2.0	2.0
1	Lox Tank Condenser Coil	10.0	10.0
3	Pressure Sensor	5.0	15.0
1	Pressure Regulator	25.0	25.0
2	Pressure Relief Valves	5.9	11.8
60	cm of 3.8mm aluminum tubing	0.2	13.2
4	Attitude Control Vanes	0.3	1.0
4	Misc Attitude Control Vane Hardware	1.0	4.0
1	Servo Tray (15.25cm x 6cm)	25.0	25.0
1	Optical GNC System	60.0	60.0
2	Button Batteries 3.6v	12.4	24.8
		TOTAL	584.6
		Budget	624.0
		Remaining	39.4

This diagram shows the basic plumbing layout of this stage. Like the earlier stage, liquid nitrogen immersed in the liquid oxygen is used as a pressurant.

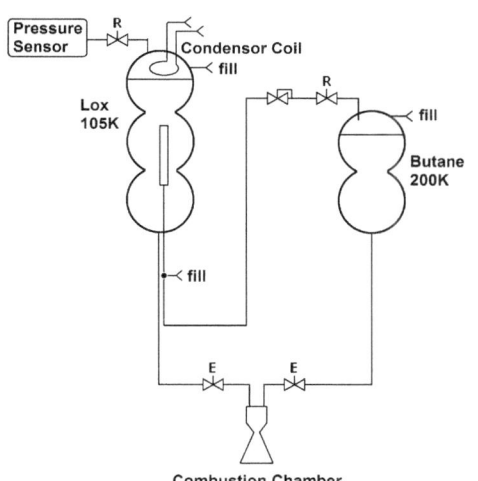

Combustion Chamber

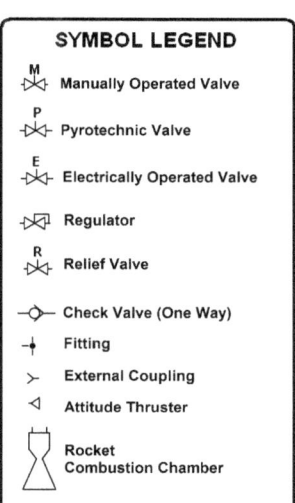

SYMBOL LEGEND

M ⋈	Manually Operated Valve
P ⋈	Pyrotechnic Valve
E ⋈	Electrically Operated Valve
⋈	Regulator
R ⋈	Relief Valve
─◇─	Check Valve (One Way)
─┼	Fitting
⟩	External Coupling
◁	Attitude Thruster
	Rocket Combustion Chamber

Propellant Tanks

The propellant tanks are constructed using electroforming from multiple joined spherical sections. The reason for this is to reduce the diameter (and hoop stress) that would be experienced with a single larger spherical shell for each propellant. The expected tank operating pressure is 415 kPa; with a 1.5 safety factor, the tanks are designed for a maximum internal pressure of 621 kPa. With an 18cm diameter, the hoop stress at maximum pressure will be about 621 MPa; this is easily within the tensile yield strength capability of electroformed nickel tanks when the thickness is about 45 microns thick. These are very thin-shelled tanks but when pressurized, they will have additional structural strength.

The liquid oxygen (LOX) tank is constructed from three 18cm thin-shelled spheroids with coupling connections in the skin. These connections can be made into the mandrel that is used to form the tanks and therefore are an integral part of the tank. In order to pressurize the liquid oxygen tank, the natural vapor pressure of the liquid oxygen is used by maintaining it at a temperature of 105 K. To ensure that the lox tank maintains the proper temperature and pressure, a condenser coil is built into it to control the temperature of the gaseous oxygen vapor above the liquid oxygen. This allows ground support equipment to monitor and adjust the liquid oxygen temperature before launching by flowing a colder fluid through the condenser coils.

Another small tank is built into the liquid oxygen tank to hold liquid nitrogen. When liquid nitrogen is held at 105K, which it would be when the lox tank is full, the vapor pressure of the nitrogen will be about 1.1 MPa. A small regulator will lower the nitrogen gas down to about 415 kPa for pressurizing the fuel tank.

The fuel tank will be constructed from two 18cm thin-shelled spheroids with coupling connections as well. The butane fuel will be kept at a temperature of 200 K in order to densify it and minimize the volume required. The rocket engine may be embedded into the fuel tank to shorten the length of the vehicle. Propellant and pressurant feed lines will penetrate the propellant tank shell in order to allow proper flows of these fluids.

Rocket Engine

The rocket engine has a combustion chamber pressure of about 206 kPa and produces about 118 Newtons of thrust. Its nozzle has an expansion ratio of about 55:1 providing the necessary Isp of 320 seconds. Like the propellant tanks, the engine can be almost completely constructed using electroformed nickel. The wall thickness at the specified chamber pressure need only be about 45 microns thick and cooling channels can be built into the combustion chamber and nozzle walls using electroforming.

The low chamber pressure has an important benefit in that the heat flux will be significantly lower than a higher pressure engine. This will facilitate the use of regenerative cooling and minimize the need for film cooling.

Even the injector cap and injector face can be produced using electroforming. This will make a very light rocket engine with high performance.

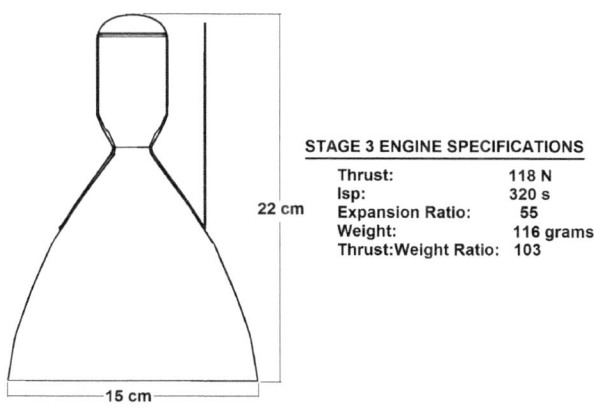

STAGE 3 ENGINE SPECIFICATIONS

Thrust:	118 N
Isp:	320 s
Expansion Ratio:	55
Weight:	116 grams
Thrust:Weight Ratio:	103

Control System

A unique control system is recommended for this vehicle: the use of control paddles to control roll, pitch and yaw. This is not a new technology, being used first by Robert H. Goddard and explained in his patent 2395809. This entails using paddles that extend the nozzle length and which swing into and out of the exhaust stream to create reaction forces. Through the use of micro servos commonly used by hobby airplane enthusiasts, the various vanes and valves can be controlled while maintaining the mass budget.

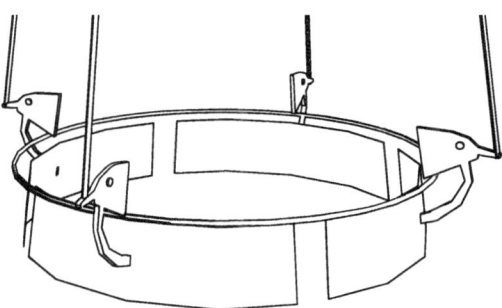

With careful implementation, it should be possible to generate roll, pitch and yaw control with a minimum loss in I_{sp}.

Chapter 10 - Booster Stage Design

In the last chapter, we looked at ideas related to upper stage design. Microlaunchers, because of their size, require some different approaches in the design of booster stages. In this chapter, ideas related to first stage boosters on microlaunchers will be detailed.

Microlauncher Ascent Profile

For both orbital and escape missions, it is advocated that a microlauncher booster take a vertical (or near vertical) trajectory in order to minimize aerodynamic forces. In contrast to other types of boosters, the purpose of a microlauncher's first stage is to lift the upper stages into close-enough of a vacuum environment that the upper stages can be designed solely for those conditions. Targeted staging altitudes for a microlauncher first stage are expected to be above 60 km.

It is worth reviewing the expected environmental conditions that the booster stage encounters. The following table shows these expected conditions through flight.

Time (seconds)	Altitude (km)	Velocity (m/s)	Air Pressure (kPa)	Air Density (kg/m^3)	Drag Force (N)	Note
0.00	0.00	0.00	101	1.2250	0	Takeoff
25.0	4.00	638	61.7	0.819	690	Transonic
34.0	7.12	447	41.1	0.5900	1236	MaxQ
67.0	34.0	1364	0.67	0.0096	235	MECO
126.0	100.0	825	0.0000	0.0000	0.05	Enter Space
210.0	134.5	0.00	0.0000	0.0000	0.00	Peak Altitude

At takeoff, the altitude, velocity and drag forces are zero. The air pressure and density is normal sea-level conditions. The rocket is expected to accelerate starting at about a 1 g upward rate until, at about 25 seconds and 4 km altitude, the vehicle passes the speed of sound. This is an important point for launch vehicles because aerodynamic forces can change drastically from what one is used to.

At about 34 seconds and 7 kilometers, the rocket encounters what is known as "Max Q." This is the point that the maximum amount of dynamic pressure is encountered. As the table shows for this example, the drag forces due to dynamic pressure are about 1.2 kN; however, it is possible for these forces to be several times larger and a launch vehicle designer may have to reduce thrust before this point is reached in order to keep these dynamic forces manageable. Often, somewhere between going transonic and the Max Q condition is where many rockets experience structural failure. A microlauncher designer must be very aware of the conditions his vehicle will face during this time period.

Once the Max Q condition is passed, the aerodynamic forces on the microlauncher will slowly be reduced towards zero. In this example, the condition known as "MECO," or Main Engine Cut Off" is reached. This is the point where the propellants have been fully consumed and the vehicle begins coasting towards its ballistic peak altitude. During this period, following MECO and where the ballistic peak is reached, the atmospheric forces continue to decrease. At some point, atmospheric forces will be reduced to a point where even large aerodynamic wings provide little or no correcting force and the vehicle may tumble.

At some point, the vehicle will reach the officially recognized boundary of space, 100 km. It is highly likely that there will be no aerodynamic forces at this point and, again, the vehicle may tumble.

Finally, the vehicle reaches its peak altitude and begins falling back towards Earth. As it falls, the aerodynamic forces will finally become significant enough that aerodynamic surfaces (i.e. wings or fins) become effective again and the vehicle will assume a nose-first attitude to the wind stream.

The potential to tumble at altitude is a significant condition to plan for. Normally, on this kind of trajectory, the vehicle may not have aerodynamic orienting forces for upwards of 200 seconds (100 seconds towards peak and 100 seconds afterward). Therefore, during the staging activities, it may be necessary to provide some form of attitude control. This will most likely be in the form of cold-gas attitude jets.

These cold gas attitude control jets need not provide control during the full period of loss of atmospheric forces, but may only be necessary at the time of staging. Therefore, the vehicle can orient itself appropriately to perform staging, perform the staging and then return to an uncontrolled condition after staging. This duration might only be on the order to 10 to 20 seconds of cold gas control.

Sources of Structural Materials

Even though microlauncher booster stages are far larger than their upper stages, there are still issues with finding appropriate materials for their construction. It is a very difficult task to manufacture every component of such things as tanks. Welding sheet into suitable tubing can be arduous and difficult. Large diameter, thin walled tubing is usually not available at most metal supply centers.

Aluminum irrigation tubing is one possible source for tubing suitable for microlauncher booster stage tanks. Aluminum irrigation tubing is often made of Al3004-H26 material and therefore demonstrates a tensile yield strength approaching 193 MPa. The following table illustrates common dimensions of aluminum agricultural irrigation tubing as well as their expected burst pressures.

Diameter (mm)	Thickness (mm)	Expected Maximum Burst Pressure (MPa)	Burst with 1.5 Safety Factor (MPa)
152.4	1.29	3.27	2.18
152.4	1.47	3.72	2.48
152.4	2.11	5.35	3.56
203.2	1.29	2.45	1.63
203.2	1.62	3.08	2.05
203.2	1.82	3.46	2.31
228.6	1.29	2.18	1.45
254.0	1.29	1.96	1.31
254.0	1.62	2.46	1.64
254.0	2.38	3.62	2.41
304.8	1.62	2.05	1.37
304.8	2.29	2.90	1.93

This wide variety of diameters and thicknesses is in a range suitable for producing tanks for microlauncher boosters.

Endcaps for Booster Tanks

Another difficult component to find is endcaps for propellant tanks. Even presuming one can obtain the tubing to make the propellant tanks, one must derive a source for suitable endcaps. It is possible to contract with a company to manufacture them in the material and sizes needed, but there is an alternative approach to manufacturing them that can be done in a shop. That method is hydroforming.

Perhaps the simplest endcaps that could be shop made are double-diameter sectors. This is a sector of a sphere that has a diameter twice the diameter of the equivalent cylinder. The reason for this odd selection of dimensions has to do with the realities of the hoop stress equations. If we ask "what diameter spherical tank would have the same hoop stress as a cylinder with the same thickness?" The answer is that the sphere has two times the diameter of the cylinder.

$$h = \frac{(2 - \sqrt{3})}{2} d$$

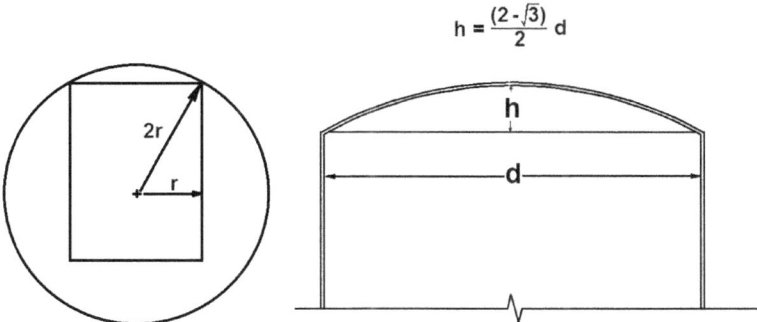

By making the endcaps with these necessary dimensions, the thickness can be the same as the cylindrical portion and it will have the same stress.

These can be produced in a simple fixture and with a hydraulic pump. The following diagram shows the principle of the fixture.

High Pressure Fluid

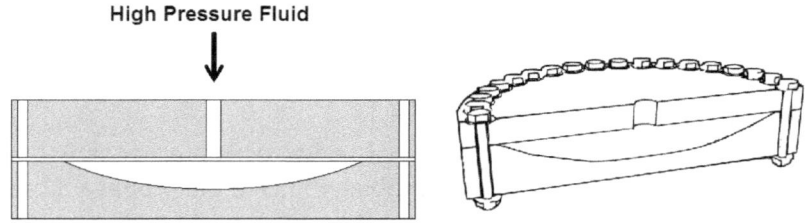

A curved cavity is produced in the bottom plate and a flat upper plate with a hole to inject high pressure fluid is clamped on top. By sandwiching a flat piece of aluminum between these two plates and injecting high pressure fluid, the flat plate will assume the shape of the curve and make suitable endcaps.

Although it is possible to weld these endcaps directly to the tank tubing, it will be better to use an intermediary assembly composed of endcap holders welded to the tanks and then the endcaps can be welded to the holders. The holders will allow the tanks to be connected by clamps and reduce the weakening effects of welds. The cutaway below illustrates this concept.

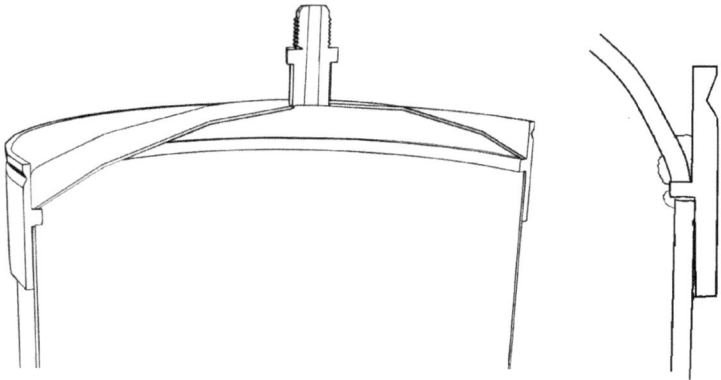

Because a slightly thicker band exists at the weld joints, the weaknesses caused by welding will be mitigated with thicker material. Additionally, a clamp can use the extended lip to connect the tank coupling.

First Stage Pressurization Systems

Although traditional pressure tank pressure fed systems can be used with first stages of microlaunchers, there are options that provide improved performance and less mass. Self pressurization and propellant vaporizers offer significant advantages over basic pressure tank pressure fed systems.

Self Pressurizing Propellants

The ability to use the vapor pressure of the propellants to feed themselves is a marked improvement where it can be utilized. By careful control of the temperatures of the propellants, the vapor pressure is controlled. Some propellants such as liquid oxygen have a useful vapor pressure at their delivered temperatures. At 90K, liquid oxygen has a vapor pressure of 1 atmosphere; at 118K, it has a vapor pressure of about 10 atmospheres. Similarly propane and butane have useful vapor pressures.

In order to best utilize self-pressurization, some ground support equipment to condition the propellants is likely needed. This allows the temperatures of the propellants to be raised or lowered as necessary quickly enough to facilitate launch schedules.

Pressurant Vaporizers

Pressurant vaporizers are another useful technique with significant mass performance improvements over standard pressure tank feeding. Of course, the downside is increased complexity, but there is sufficient enough of a benefit to outweigh this downside. Although details of this technology are to be worked out, we can sketch out the principles behind some of these mechanisms.

Pressurant vaporizers utilize a heat source which is used to vaporize the pressurant sufficiently to provide pressure for the propellant tank ullage space. In one approach, some of the propellant is extracted from the tanks, burned and the heat is transferred through a heat exchanger to vaporize a propellant. In another approach, some of the propellants is extracted from the tanks and burned to vaporize a separate pressurant gas, for example, liquid nitrogen.

Despite the complexities of these vaporizing systems, the mass saving can be significant over pressure tank pressure fed systems.

Staging Mechanisms: The Marman Clamps

There are many different ways to reliably perform staging separation needed by multistage launch vehicles. The Marman clamp offers one easy, reliable, and often-used technique to do it. The Marman clamp is a kind of band clamp that can apply significant holding force yet which is easily released. The Marman clamp might actually be considered to be composed of three parts. There is one coupler piece which is attached to one of a to-be mated stage, another coupler piece that is attached to the other to-be mated stage and a steel band with a release mechanism.

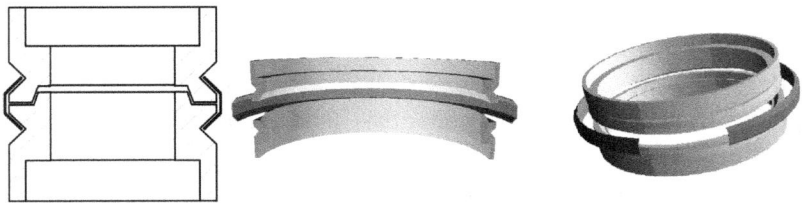

Through an adequate release mechanism, the band can be made to reliably and quickly release the two stage components from each other. Although some applications use exploding bolts as a releasing mechanism, it is also possible to use actuated pins to release in a non-pyrotechnic manner. It is also possible to apply the same principle inside of the body of the vehicle so that the part of the body that meets the airstream is smooth and all staging mechanisms are in the interior of the vehicle.

 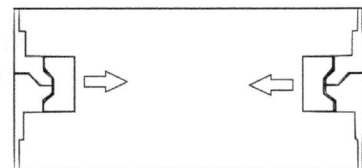

An Example First Stage: ML-1

The following is a detailed description of a first stage design suitable for a microlauncher. It can carry sufficient enough payload that upper stages can be carried to their operational altitudes. It is based on self-pressurizing propellants: liquid oxygen and propane, and utilizes 25.4 cm diameter irrigation tubing tanks.

System Specification

Overall, the vehicle is a three stage launcher with the upper stages similar to what was described in the last chapter. The first stage is capable of carrying these payloads to altitudes approaching 121 km. It utilizes a 3.34 kN main engine with two small 100 N vernier engines to provide roll, pitch and yaw control throughout the entire flight up through staging. Four fins provide additional aerodynamic control during ascent.

This first stage is a relatively low-technology vehicle for high-technology upper stages. A first stage vehicle like this, using advanced electroformed upper stages could be expected to put about 200 milligrams onto an escape trajectory. Its sole job is to raise the upper stages to where they can operate in vacuum-like conditions. No recovery budget is allocated so this is a single flight vehicle. But, being a low -technology vehicle, it is expected that its construction costs will be relatively low.

Because of its relatively small size, it is expected that microlauncher enthusiasts can develop this vehicle stage through repeated test flights at relatively low cost. After the first stage has been developed and perfected separate from its payloads, different degrees of upper stages can be tested and perfected in expanded envelope test steps.

The following table details the rocket equation based performance requirements and specifications of the first stage.

Parameter	Value	Units
Payload	32.693	kg
Oxidizer	Liquid Oxygen	
Fuel	Propane	
Average Flight I_{sp}	249	Seconds
Desired dV	2438	m/s
Total Propellant Mass	115.407	kg
Oxidizer Mass	79.342	kg
Fuel Mass	36.065	kg
Ms	34.622	kg
Me	67.315	kg
Mf	182.723	kg
Diameter	25.4	cm
Length	7.2	m
Main Engine Thrust	3.34	kN
Vernier Engine Thrust (each)	100	N

With the specified performance, the vehicle is expected to reach a maximum altitude of about 121 km. The following table illustrates the mass budget.

Qty	Description	Unit Weight (kg)	Extended Weight (kg)
1	Propellant Tanks	9.26	9.3
1	Payload Bay	6.77	6.8
1	Nosecone	1.28	1.3
1	3.34 kN Main Engine	3.18	3.2
1	Engine Gimbal System	3.56	3.6
2	100 N Roll Engines	0.36	0.7
1	Battery	0.23	0.2
2	Throttle Servo Valves	0.91	1.8
4	Fill/Drain Valves	0.11	0.5
2	Pressure Sensors	0.45	0.9
1	Fin Can	3.31	3.3
	TOTAL		31.5
	Budget		34.6
	Remaining		3.1

Propellant Tanks

The propellant tanks are based on 25.4 cm diameter, 1.29mm thick irrigation tubing. The oxidizer tank has a volume of 73 liters, providing about 5% ullage volume. The fuel tank has a volume of 70.4 liters which has a 5% ullage as well. Based on these volumes, the cylindrical portions of the tanks are 1.44 meters and 1.39 meters respectively. The tanks are constructed as described earlier and as the following drawing shows.

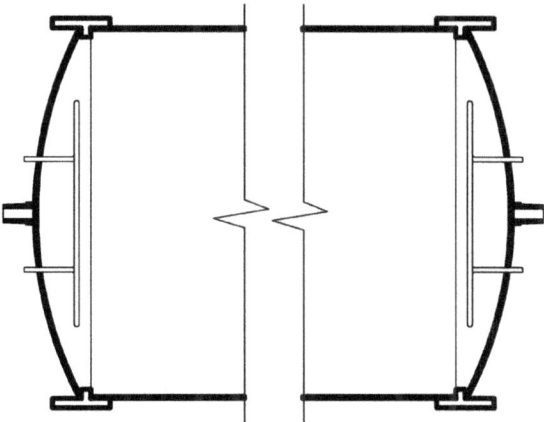

The tank plumbing is somewhat simpler because of the simplicity of self-pressurized propellants but includes heat exchanger coils inside the top and bottom of each tank. These coils allow heating or cooling fluids to be circulated inside of the tank to facilitate conditioning the temperature of the propellants (and thus their pressures).

If necessary, small propellant mixing paddles could be included in the tanks to mix the propellants and produce a consistent bulk temperature during the propellants' conditioning.

The main engine is fueled from liquid drawn from the bottom of each tank while the vernier engines are fueled from vapor drawn from the top of each tank.

The following drawing shows the plumbing diagram for the propellant tanks and propulsion system.

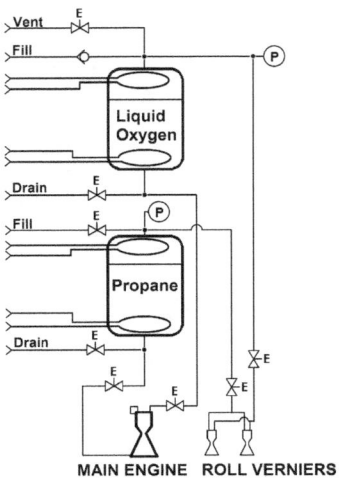

MAIN ENGINE ROLL VERNIERS

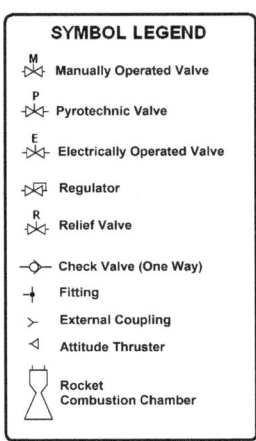

To condition the liquid oxygen propellant, the tank would first be filled with 90K liquid oxygen. To raise the temperature of the liquid oxygen in the tank, room temperature propane would flow through the lower heat exchanger. This would cause the temperature of the liquid oxygen to increase. During launch delays and while performing propellant conditioning on other tanks, additional 90K liquid oxygen would be circulated through the upper heat exchanger in the liquid oxygen tank. This would cool the liquid oxygen. Motorized paddles may be necessary to mix the propellants to homogenous temperatures.

To condition the propane, its tank would first be filled with ambient temperature propane. It would most likely need to be heated to increase its pressure. Warm water would flow through the lower heat exchanger in the propane tank to warm the propane. As needed, the propane could be cooled using cold water through the upper heat exchanger.

Pressure sensors would be used to monitor the state of the propellants and their conditioning process.

Payload Bay and Nosecone

The payload bay would be constructed of the same material as the tanks and would be approximately 2.4 meters in length. This would provide sufficient volume for carrying upper stages. As a basically empty tube with the upper stages inside, the upper stages can fly out of the tube at the staging altitude.

The nosecone could be made from a composite material or aluminum.

Main Engine

The main rocket engine is a 3.34 kN regeneratively cooled engine which is gimbaled in two axes. Operating with a chamber pressure of about 10 atmospheres and with an expansion ratio of about 3.5:1, it would provide an I_{sp} of approximately 212 s at sea level and 276 s in a vacuum for an average flight I_{sp} over 250 s; it would operate for over 70 seconds.

Parameter	Value	Units
Thrust	3.34	kN
Chamber Pressure	10	atm
Expansion Ratio	3.5	
I_{sp} (SL)	212	s
I_{sp} (Vac)	276	s
Throat Diameter	5.96	cm
Exit Diameter	11.16	cm
OF Ratio	2.2	
L*	51	cm

There are many different ways to build the engine with these parameters. The following diagram shows one way which is based on commonly available tubing with a minimum of machined parts. Almost all of this can be constructed on a lathe.

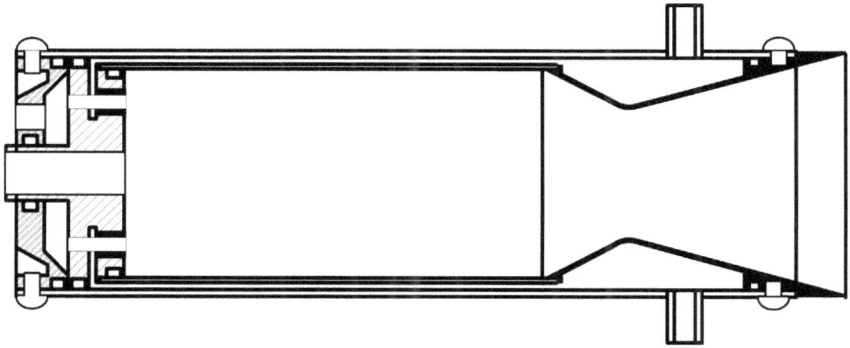

By careful selection of off-the-shelf tubing, it is possible to construct a workable engine combustion chamber and coolant sleeve. The nozzle section is machined from steel and the injector assembly is made of aluminum.

Vernier Engines

There are two 100 N vernier engines to provide roll control during main engine firing and then full roll, pitch and yaw control after the main engine is shut down. They are operated from the ullage gases of the propellants to allow good performance in small packages. These engines operate at 5 atmospheres with an expansion ratio of 1.5.

The following table gives the characteristics of these engines.

Parameter	Value	Units
Thrust	100	N
Chamber Pressure	5	atm
Expansion Ratio	1.5	
I_{sp} (SL)	185	s
I_{sp} (Vac)	240	s
Throat Diameter	1.47	cm
Exit Diameter	1.80	cm
OF Ratio	2.2	
L*	51	cm

It is highly likely that these vernier engines can be ablatively cooled. This will simplify their construction use and increase their reliability. The following diagram shows what these small vernier engines might look like.

This engine uses a graphite nozzle and an ablative phenolic liner cooled by the incoming gaseous fuel to allow the long-duration burn needed.

Engine Gimbal

The engine gimbal allows the main engine to be directed to ensure the

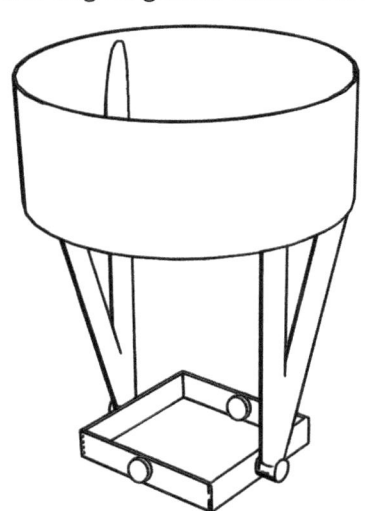

rocket is following the proper trajectory. It is constructed from aluminum with bearings at the rotation points. It also conveys the engine forces to the tanks in a way that does not cause structural failure. Most often, the gimbal is connected to the head of engines, but this is not the case in this design. The gimbal of the ML-1 vehicle, in order to save space, is connected to the main engine at the nozzle end of the engine.

Another unique aspect about the gimbal arrangement is the use of individual hinges for each vernier which are attached to the main engine. This is shown below.

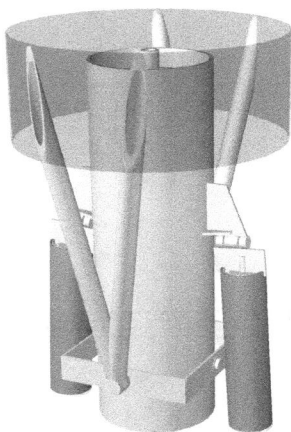

With this arrangement, the engine can provide full roll, pitch and yaw control throughout the entire flight while fitting into the body diameter. During the main engine burn, the main engine provides yaw and pitch control and the vernier engines provide roll control. After the main engine is shut down, the vernier engines provide full roll, pitch and yaw control.

Flight Profile

The reason for the unique design of the gimbal and engine arrangement is to support the necessary flight profile and control during the flight profile. The flight profile entails using the main engine to provide its full 3.34 kN thrust while the vernier engines control roll. At about 3.2 km altitude, the main engine is throttled back during the burn to ensure that the dynamic pressure faced by the vehicle does not exceed a desired level. As the dynamic pressure decays and the vehicle accelerates, the main engine is throttled back up to near full thrust. At about 32 km altitude, the main engine is shut off and the vernier engines continue burning to provide full roll, pitch and yaw control up to the staging point at better than 65 km. During the vernier-only burn time, the vernier engines provide sufficient thrust to ensure that there is always at least 1/10th of a gravity acceleration. This profile is necessary because the vehicle must be fully controlled while releasing the upper stages and the upper stage propellants must be settled to ensure proper staging.

Propellant Handling and Conditioning Equipment

All of the microlauncher vehicles shown in this book have used liquid oxygen as the oxidizer and either butane or propane as the fuel. In many of these cases, it is necessary that these propellants be at particular temperatures (or not exceed certain temperatures) to ensure correct and safe operation. Since it is possible that a rocket launch may be delayed due to a number of factors, it is necessary to bring the propellants to their proper operating temperatures and maintain that temperature until launch. This is the job of propellant conditioning equipment.

Propellant conditioning equipment may need to raise or lower the temperatures of the propellants and, therefore, the capability to increase temperatures and lower temperatures must be included in the design of propellant conditioning equipment.

Liquid Oxygen Dewars

When delivered, liquid oxygen will most likely arrive in a dewar. A dewar is a large vacuum flask which uses an evacuated chamber around the contained fluid to act as an insulator. Because a vacuum is such an excellent insulator, liquid oxygen can remain in the dewar in chilled, liquid form for many hours to days.

Since the cryogenic liquids held in dewars eventually boils to vapor, dewars are fitted with safety relief valves to relieve the built-up pressure. This build up of internal pressure is also used to feed the liquid products out of the dewar. You should take some time to learn how to safely handle liquid oxygen and its safe storage in dewars.

Common dewar sizes as would be used for first stages of vehicles are 160 liters, 180 liters, and 250 liters. Each of these is far more liquid oxygen than is normally used for microlauncher vehicles. However, it is convenient to receive such larger quantities since some of the liquid oxygen will evaporate in handling, some will go into the vehicles and the rest may be used to condition the propellants' temperatures and pressures.

An Example Propellant Handling System

For the ML-1 vehicle, it is necessary to both heat and cool the liquid oxygen and heat and cool the liquid propane to ensure proper and safe storage and operation of these propellants. A suitable propellant handling system would consist of a supply of liquid oxygen in a dewar plus liquid water which can be heated and cooled to condition propellants such as butane and propane.

If you recall from the last sections, heating and cooling coils were included in the propellant tanks of microlauncher vehicle stages. This technique readily allows conditioning of the propellants. A suitable propellant conditioning system would be able to supply either liquid oxygen or water under pressure at the correct temperatures as is necessary. These fluid sources would be connected to the vehicle on the launch pad or tower and measure temperature and/or pressure to automatically condition the propellants.

In the case of the ML-1 first stage vehicle, it was necessary to maintain the pressure of the propane at about 15 atmospheres. In order to do this, propane must be kept at a temperature of 315 K or 42° C. This temperature is within the range of warm or cool water to be used as a conditioner. Therefore, ice-chilled water at about 4° C can be circulated in the upper vapor coil of the propane tank to cause the temperature of that propellant to be cooled. Warm water at about 65° C can be circulated in the lower liquid warming coil to cause the temperature and pressure of the propane to be increased.

In the case of the ML-1 vehicle's lox, liquid oxygen at about 90 K from a dewar can be circulated in the upper vapor coil of the lox tank to cool that propellant and room-temperature propane gas can be circulated in the lower liquid warming coil to cause the lox to warm up. In this manner, the temperature and pressure of the liquid oxygen can be maintained for long durations of time.

Therefore, a complete propellant handling and conditioning system would be able to deliver two propellants and circulate heating and cooling fluids as necessary to maintain the proper temperatures and/or pressures. The following diagram shows a basic schematic of how one might appear.

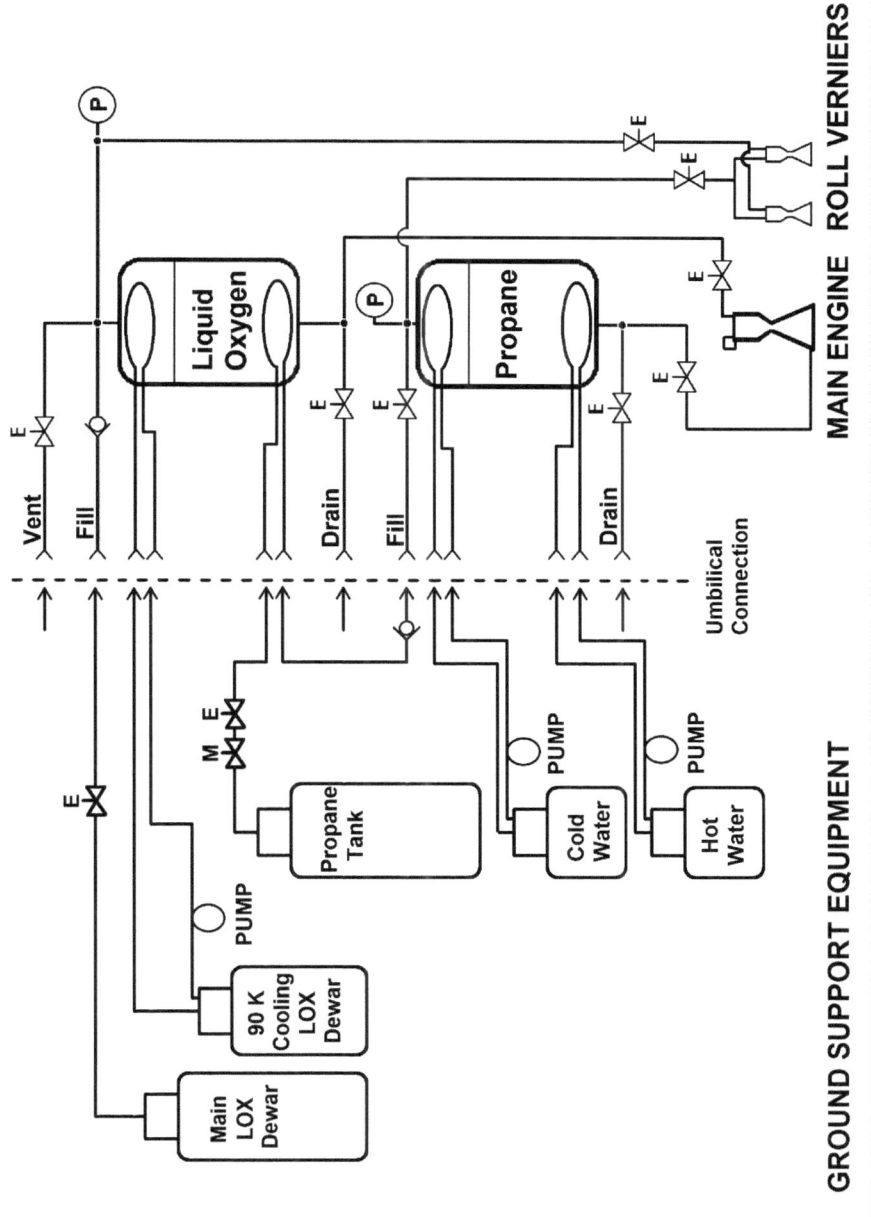

ROLL VERNIERS

MAIN ENGINE

Liquid Oxygen

Propane

Vent

Fill

Drain

Fill

Drain

Umbilical Connection

PUMP

Main LOX Dewar

90 K Cooling LOX Dewar

Propane Tank

Cold Water

Hot Water

PUMP

PUMP

PUMP

GROUND SUPPORT EQUIPMENT

Chapter 11- Microlauncher Spacecraft

Even though the emphasis so far has been on rockets, the reason that microlaunchers exist is to carry small spacecraft beyond Earth's atmosphere. We will briefly discuss important aspects of small spacecraft suitable for microlaunchers and their methods of sensing and communicating.

The first and most obvious characteristic of microlauncher spacecraft is their small size. Of the vehicles detailed in this book, most of them have a payload capacity of near 150 to 200 grams to an escape trajectory. This payload limitation imposes unique requirements on microlauncher spacecraft and dictates imaginative solutions to overcoming the problem of getting sufficient functionality in a small, lightweight spacecraft.

But a fully functioning microlauncher spacecraft is possible. One good reference source for what is capable for small spacecraft is the modern smartphone. These are amazing demonstrations of technical sophistication in small packages. Inside of modern smartphones, one finds powerful computers with significant amounts of long term memory, high resolution video and still cameras with sophisticated lens systems, GPS localization systems, inertial sensors like accelerometers and gyros, WiFi and phone radio systems, temperature sensors, vibration and audio sensors, battery and power management systems and many other kinds of sensors and actuators. All of these capabilities are packaged in masses and sizes that would make ideal spacecraft control systems. Of course, there are also features that are lacking, but the size and mass of smartphones represents the possible capabilities for a microlauncher spacecraft.

The Spacecraft Bus

To simplify the development of microlaunchers spacecraft, it is recommended that a generic bus be designed and developed. A spacecraft bus is a general non mission-specific platform onto which mission-specific enhancements are attached. The spacecraft bus contains the bare minimum common subset of subsystems which many conceivable missions would require. Typical components included in a spacecraft bus would be:

- Structural Components
- Computer System
- Electrical Power System
- Thermal Control
- Attitude Control System
- Guidance, Navigation and Control System
- Propulsion System
- Communication System

By having a general purpose spacecraft bus which has a large subset of these features, developing the complete spacecraft for any given mission becomes much easier. One merely starts with the bus and adds the mission-specific sensors and other capabilities that are needed. The following paragraphs describe the bus components in greater detail.

The Structural System binds, protects and aligns all of the elements of the spacecraft together. Like the keel and hull of a ship, the structural system provides the strength and shape of a spacecraft.

The Computer System of a spacecraft performs multiple functions. It provides the timing and sequence control, it collects, manages and disseminates the system's information, it controls and coordinates all of the spacecraft subsystems to perform the mission of the craft.

The Electrical Power System collects, stores, manages and distributes the electrical power needed to operate all of the subsystems. The system may include solar cells or other power generating devices like reactors. Batteries are often used to store the electrical energy for use when external power sources like solar light are not available. It may also convert, switch and distribute electrical power to devices as needed.

The Thermal Management System manages the heat conditions of the spacecraft. A spacecraft has two opposing problems that it must manage: an abundance of heat and a steady loss of heat. Heat sources like the sun provide large heat fluxes which must be protected against, while the vacuum of space acts like an insulator to keep heat in. However, the blackness of space also allows the heat to radiate away on dark sides. All of this must be managed to guard against harmful temperatures.

The Attitude Control System ensures that the spacecraft is pointed in the correct way. It may do this via active and passive means. Active mechanisms such as rocket thrusters and reaction wheels represent some of the ways to control the attitude of the spacecraft. Additionally, passive techniques using sunlight radiation pressure in the manner of solar sails.

The Guidance, Navigation and Control System (also known as the GNC system) causes the spacecraft to follow or assume trajectories that it must be on for its mission. It determines where the spacecraft is, where it needs to be and performs the operations needed to make it go where it's supposed to be.

The Propulsion System is used to control the speed and trajectory of the vehicle. There are active and passive propulsion systems. Active systems might use rockets whereas passive systems might use solar sails or other techniques of propulsion.

The Communications System provides a means to communicate between Earth and the spacecraft (or between spacecraft).

All of these subsystems work together to provide support for mission-specific spacecraft payloads to perform their job.

Mission Specific Spacecraft Systems

Whereas the spacecraft bus consists of all the non-mission specific elements of the spacecraft, the spacecraft bus can be modified to be suitable for a complete mission merely by adding additional subsystems. Mission specific systems might include sensors like telescope and radiation sensors or additional computing or data storage systems.

A Budget Based Analysis

One of the first ways to start looking at the functional requirements of a spacecraft bus is to assign a mass budget based on other designs and then scale it down as a predictor of sizes and masses needed in the microlauncher spacecraft bus.

The following table shows an example mass budget based on a real spacecraft and then shows it scaled to provide a preliminary budget for a microlauncher spacecraft.

Subsystem	Percentage	200 gram Microlauncher Spacecraft Example Budget
Mission Sensors	16.0%	32.0 grams
Power	18.4%	36.8 grams
Communications	7.1%	14.2 grams
GNC	9.7%	19.4 grams
Computer	5.3%	10.7 grams
Structure	29.2%	58.4 grams
Thermal Control	3.1%	6.3 grams
Wiring and Interconnect	11.1%	22.2 grams

Using these mass examples, we can begin to appreciate the scale of the task in developing these subsystems. For example, the structure subsystem is budgeted as 58.4 grams; a typical empty soda can weighs about 15 grams. Through careful design and integration of these various systems, components may be able to perform multiple functions and allow much more capability. For example, the communication function might be merged with the sensor function to allow their combined mass. This might occur, for example, by allowing a telescope component to function for both communications and mission specific photography.

Laser Communications

Optical communications is a very capable and efficient means for small spacecraft to communicate bidirectionally with Earth. Sensors like CCD's, Photomultiplier Tubes (PMT's) and Avalanche Photodiodes (APD's) can be sensitive enough to discriminate signals consisting of only several thousand photons when adequately filtered. Small laser diodes in the range of 30 milliWatts can provide highly directional photon sources that can be detected over a couple of million kilometers using these kinds of sensors. Coupled with a small telescope, this creates a complete lightweight long-range spacecraft communication node which can be monitored from Earth. Even simple on-off keying (OOK) modulation can provide sufficient signal-to-noise ratios to be used over long distances.

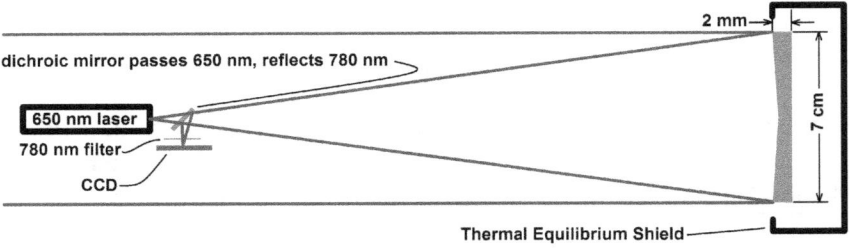

In this example integrated transceiver, a small silicon mirror is used for both transmitting and receiving signals. A dichroic mirror is used to separate the two different laser wavelengths used for transmission and reception.

Cansats

A Cansat is a specification for a small soda can size spacecraft. The Cansat specification calls for the spacecraft to be the size of a soda can, 66mm in diameter by 115mm in height, and to have a mass below 350 grams. The approach of using a soda can as a container demonstrates a potential packaging method for microlauncher spacecraft. Many Cansats have flown on sounding rockets as educational projects.

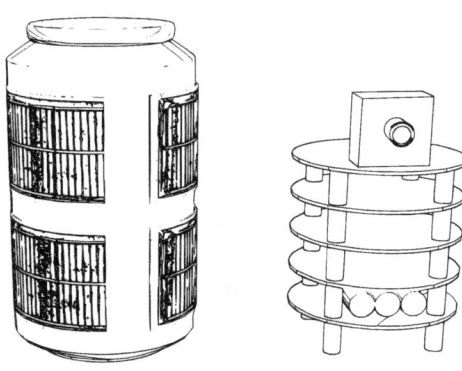

There have been some very sophisticated CanSats over the years and it shows the potential for non-professionals to develop spacecraft that could fit on microlaunchers. There have been CanSats with sophisticated computers, video imaging cameras, bidirectional radios, GPS receivers, atmospheric sensors, inertial measurement units and sophisticated servo controlled mechanisms. There have even been CanSats with reaction wheels to control rotation and orientation.

Even though no CanSats have flown to space, there is innovation, adventure, excitement and a spirit of experimental development which represents what is envisioned for microlauncher spacecraft. Rather than being hard core, multiyear aerospace projects, it is hoped that microlaunchers enable the kind of experimental development seen by CanSat developers. CanSats are developed quickly over a period less than a year, are hand loaded onto their sounding rockets, and launched. Microlauncher spacecraft are envisioned as developing in much the same way.

An Example Microlauncher Spacecraft Bus

Now that we have an idea of the required subsystems, their functions and allocated masses, we can begin to consider what a microlauncher spacecraft bus might look like. The following image shows different views of one recommended design.

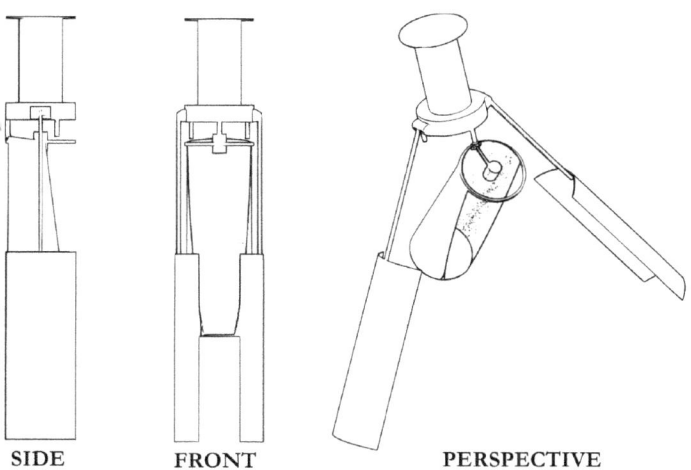

SIDE   FRONT   PERSPECTIVE

This example bus is a basic assembly able to support various missions. Its main functions are to provide power for the spacecraft, maintain electronic equipment temperatures, maintain the spacecraft's orientation, maintains a laser diode data link, and support instruments for experiments. Major components are identified in the following diagram.

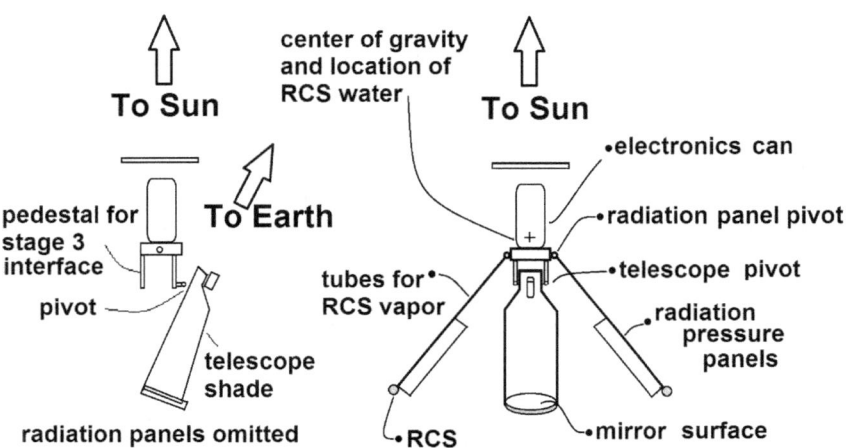

This spacecraft bus depends on using solar radiation pressure for long-term orientation maintenance, a circular solar panel for power generation and instrumentation shading, and a telescope for laser communications and imaging. It also provides an enclosure for a computer board and other instrumentation.

A short -term use reaction control system is envisioned to provide initial orientation of the vehicle after launch. It likely uses water or ammonia as a cold gas monopropellant. This system would orient the vehicle into the initial proper attitude and roll rates that would then allow the radiation paddles to take over. The use of solar pressure to control the orientation of spacecraft has been demonstrated repeatedly and is an established technique. Most recently, it has been used on the Kepler planet finding spacecraft to compensate for reaction wheel failures.

The following diagram shows a conceptual block diagram of the various functions included in the bus.

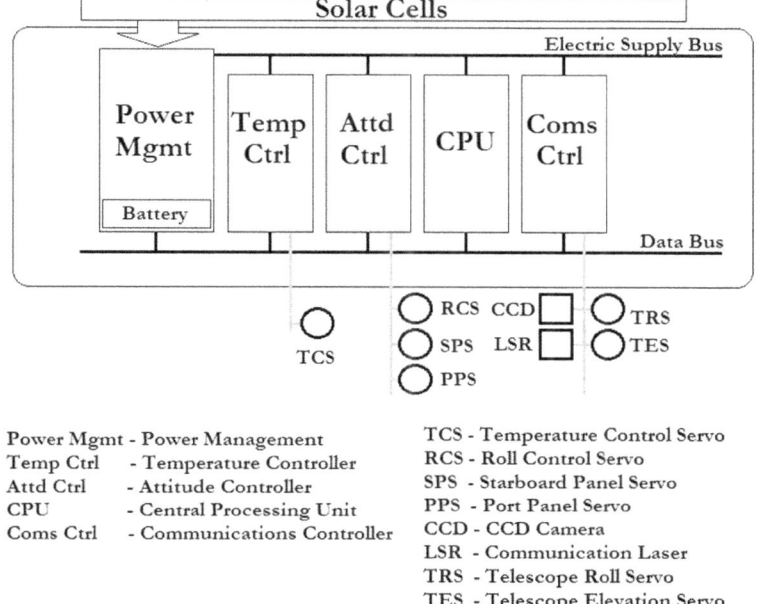

Power Mgmt - Power Management
Temp Ctrl - Temperature Controller
Attd Ctrl - Attitude Controller
CPU - Central Processing Unit
Coms Ctrl - Communications Controller

TCS - Temperature Control Servo
RCS - Roll Control Servo
SPS - Starboard Panel Servo
PPS - Port Panel Servo
CCD - CCD Camera
LSR - Communication Laser
TRS - Telescope Roll Servo
TES - Telescope Elevation Servo

Each of these subsystems will be discussed briefly here.

The Solar Cells collect light energy from the sun and convert it to electric power. They also serve as a thermal shield for the rest of the vehicle by shading the electronics can from the sun's heat when the cells are pointed directly toward the sun.

The electricity from the Solar Cells goes to the Power Management System in the electronics can. This unit regulates, stores and distributes power to the various subsystems of the spacecraft.

The Temperature Control System utilizes a temperature sensor to monitor the temperature of components in the electronics can and then activates the Temperature Control Servo to open or close mirrored/blacked panels on the body to cause heat to be absorbed or reflected from the exterior of the electronics can.

The Attitude Control System uses small gyros and optical sensors to control the radiation panels to orient the spacecraft in 3 rotational axes (roll, pitch, and yaw). An initial water-based reaction control system would be used to provide a rough initial orientation of the spacecraft after which the radiation panels take over orientation control.

The Central Processing Unit controls all other systems to cause the spacecraft to complete its mission. It has data storage and processing capabilities as well as communications to all other subsystems.

The Communications Control System operates the telescope through servos to cause it to point in the correct direction relative to the spacecraft body. It then utilizes the CCD camera and laser diode to perform communications operations. The CCD is also used to take images which are stored by the CPU.

The various servos used by the bus must be extremely small and lightweight. There are several options for these. There exist many small servos used by hobbyists which are as small as 2.5 grams but these are unlikely to survive the space environment for very long. Still, with modification, by replacing plastic parts with suitable metal parts, they may be suitable. Another option is small voice-coil actuators based on tiny neodymium magnets. These have the small size and strength necessary to do the job, although some means of position feedback may be required; Hall Effect sensors may work here.

All of the electronic components of this spacecraft bus are likely mounted onto a small circuit board inside of the electronics can. The electronics can could possibly be an actual soda can with a sea level pressure non-reactive gas like nitrogen inside. This would simplify many aspects of thermal management as well as operation of electronic circuits in space without increasing the mass significantly. Electric control signals could enter and leave the top of the can through hermetic seals.

The telescope can use a thin silicon or quartz mirror with a focal length of about half of a meter and a diameter of between 7.5 cm and 10 cm. At the focal point of the telescope would be a small assembly containing an interference filter mirror, a CCD camera chip and a laser diode. This would allow bidirectional laser communications as well as high resolution imaging. A lightweight shade would protect the mirror and sensors from the direct light of the sun.

All of these basic subsystems together constitute the spacecraft bus. By itself, it can do some useful missions without additional sensors or actuators. It can perform imaging and communications experiment missions on escape trajectories which parallel the Earth's orbit. In this manner, it would be visible in the sky for several hours per day for many months.

Your Microlaunchers Future

Chapter 12- Moving Forward

In the previous chapters, we've introduced the microlaunchers idea and the basis for this revolution. We then introduced the underlying concepts of launch vehicle technology necessary to bring about this change. Finally, we went into a bit of depth on specific technologies used in these small launch vehicles. Since what is needed for this revolution is a new approach of technological development, a lot of emphasis has been placed on the technology of launch vehicles. The reader should now have a good idea of the scope of the task associated with constructing microlaunchers and their components.

In this chapter, these ideas will be concluded with ideas on how you might be able to involve yourself in creating the microlaunchers revolution. We will introduce personal projects that you can work on to bring about the microlaunchers revolution as well business opportunities where the technology can become financially well-established. But first, a little bit will be said about the microlaunchers revolution.

A Vision of a New Revolution in Space Development

Microlauncher technologies will make space more accessible and affordable for more people. The existence of microlaunchers will smash the idea that space is only something that governments and big industries do. In much the same way that microprocessors brought computing to more people, microlaunchers will bring space to more people. Once these small vehicles exist, they will change how people think about space.

Space today is only done by big industry and government for their own purposes and the cost of launch vehicles and spacecraft are insignificant compared to the great profit and large competitive advantages these vehicles bring. These institutions have no need to lower these costs. In fact, it serves their purposes if their competitors cannot afford them as well. The profits, competitive advantage and cash flows are at such large levels that launch vehicle and spacecraft costs are in the noise for these institutions.

Microlauncher missions change the economics of space, however. Microlaunchers create new needs and desires for the average person and small groups to pursue. There is no rational reason why the cost of space access cannot become more affordable in the future. If one looks only at the material costs of microlaunchers and their spacecraft, there is at most several thousands of dollars worth of material: tens of pounds of aluminum and steel plus some electronics components that are already being used in commercial consumer electronics. Even propellant costs are insignificant in this bigger picture. But, because the average person does care about the cost, they will drive the costs down to the level that they can afford, once they see some benefit in pursuing space.

Whereas today there is a dearth of suitable rocket engines and vehicles, the microlaunchers revolution will make a abundance of rocket technology available. Just as there are hobby airplane fliers with jet engines, there will be hobbyist rocketry people who take off-the-shelf rocket engines, tanks and upper stages, and then beef them up make them more efficient and apply them to their own missions. Think about that reality: secret technology from World War II (specifically jet engines but also rockets) has become an essential part of a hobbyist level pastime in RC jet flying. It has happened for jet engines and it can happen for rocketry as well.

People will scale up and scale down this technology to suit their needs and abilities. Microlaunchers technology can be scaled up to that of existing launch vehicles. Clusters of smaller microlauncher first stages will be able to place larger payloads into space. By scaling up the engines, tanks and upper stages, larger vehicles will be developed from this foundational technology in a much more affordable and accessible way than is available today.

Universities will do their own space missions where one class builds each of the major components: one class for a first stage, one class for a second stage, one class for the third stage, one class for the spacecraft. One class might be responsible for launching these vehicles whereas a later class operates the ongoing mission. Other classes might analyze the data collected by these missions. Students who've worked on these missions will go off, start their own companies and make their expertise available through products and services.

The availability of microlaunchers and spacecraft will change the economics for industry and government as well, though. Existing government space exploration agencies might consider using these small, cost-effective vehicles to do more science with less money. Spreading swarms of small spacecraft across the solar system on drastically different trajectories will give us a contact and awareness of the solar system like has never been known. For example, measuring solar particles streaming off the sun from 100 different spatially separated points simultaneously becomes affordable at the cost of one of the "big space" missions today.

But it is in how the average person can explore the solar system that will allow microlaunchers to change the future. If someone can launch their own spacecraft on their own desired trajectory for the cost of an expensive vacation, then the average person can do their own virtual reality tours of the moon or an asteroid. Groups of individuals can set up their own data relay spacecraft and work cooperatively to do their own missions to outer planets and asteroids.

Maybe a small company can afford to put a prospector spacecraft on an asteroid and begin extracting useful and profitable minerals. At the least, they could survey asteroids and sell that information to someone who might be able to extract those minerals.

All of this becomes possible with the availability of affordable small launchers and spacecraft.

Personal Projects

There are many different ways for an individual to involve oneself in bringing about the microlaunchers revolution. We would like to review explicit ideas that might help you see how you can be involved. What follows is a brainstormed list of personal projects.

Learn about the Rocket Equation

The Rocket Equation is one of the key design equations used to specify the requirements for rocket flight. Knowing this equation is an important part for you to be able to appreciate the significance of aspects of microlauncher design. We've introduced it in this book. Study it and understand it. If you are having trouble, do further research elsewhere until you fully understand its significance.

Learn about rocket engines and combustion

Microlaunchers depend on small and efficient rocket engines. The principles of efficient combustion, the structural issues due to the high temperatures involved and the engineering of small combustion chambers able to reach space is an important discipline to learn about.

Learn about orbital mechanics

Getting small payloads into accurate orbits or on useful escape trajectories requires knowledge of the basic physics of orbital mechanics.

Learn about Aerodynamics of Small Launch Vehicles

As the first stage of a microlauncher ascends through the atmosphere, there are significant aerodynamic problems are encountered. You can set out to learn about these from many different sources.

Find and involve yourself with a Microlaunchers Interest Group

If one exists in your area, you could involve yourself with others who also want to learn about and develop microlaunchers. This way, you can involve yourself in others' projects, get others involved in your microlaunchers projects.

Start a Microlaunchers Interest Group in your area

If an interest group doesn't exist in your area, then think about starting one. You could get others involved in learning about and making small liquid propellant rockets.

Develop an igniter for a rocket engine

This is actually a good introduction to all of the parts of a full engine and in many ways is a much more manageable project for beginners. In any case, it'd be nice to have a reliable igniter before you start testing an engine.

Develop a first stage engine

You could start working on the development of your own design or use others' designs for a first stage engine. Once you have a good, working design, you might think about making it available to others to advance the microlaunchers revolution.

Develop a First Stage

A broader task is the development of an entire first stage. This will allow you to learn about and develop skills in pressurization systems, rocket engines, control systems, igniters. You will develop a lot of important skills in this way.

Develop an upper stage engine

Upper stage engines are smaller than first stage engines and might be easier to develop.

Develop an upper stage vehicle

An entire upper stage vehicle is a larger, multidisciplinary task that'll develop a lot of skills.

Develop a guidance and control system

You might consider developing the guidance and control system for either first or upper stages. There's a lot to learn about here in electronics and closed loop control.

Develop a ground control station

Once the first microlauncher spacecraft leave Earth, there will be need to communicate with them and uplink commands or downlink images and data.

Develop a Laser Comm link

Laser communication links are an exciting and useful way to communicate with spacecraft. Gain knowledge and experience about modulation techniques, laser diodes and various photon counting sensors as Charge Coupled Devices (CCD's), photomultiplier tubes (PMT's) and Avalance Photodiodes (APD's).

Develop a turbopump for a first stage engines

Although turbopumps are an advanced technology, their use will enable smaller and more efficient rocket engines.

Get involved with a launching group

Link up with others interested in creating microlaunchers and work as part of a team to be able to put a microlauncher spacecraft on an escape trajectory.

Operate a mission just launched

Once a microlauncher spacecraft has been launched, it will be important to send control commands and receive the images and data from spacecraft. Volunteer to be part of a group which is operating a microlauncher spacecraft.

Business Opportunities

Getting involved with microlauncher launch vehicles and spacecraft will cause you to learn and develop very useful skills which you might leverage into business opportunities. The systems and subsystems that you develop will be important products for others doing space science and businesses. Here are some ideas of how you might be able to leverage these skills and products into businesses.

Component Manufacturing and Sales

The various engines, valves, regulators, tanks, and control system components that you're involved with making become saleable products to others.

System Manufacturing and Sales

Once you've gotten good at making first stages and upper stages, these systems can be sold to others who merely want to assemble missions from higher-assembly parts.

Microlaunchers Training Company

The skills and knowledge that you've learned doing your own microlaunchers are valuable to those who don't know them yet. You could offer to train others in seminars and classes in your specialties.

Launch Services Company

One could use a developed launch vehicle, either as a complete set of stages, the first stage alone or with the upper stages for flying customers' experiments for them.

Lease, Wet or Dry

Launchers and support equipment, with launch crew can be leased for customer missions (this is known as a "wet lease"). Optionally, the launchers and support equipment can be leased to a trained customer launch crew (this is known as a "dry lease"). These terms are derived from aircraft leasing services. In either case, the leases would consist of both expendable and reusable components.

A Mission Operation Company

You might also offer your services and equipment for rent or sale to others. You've developed a lot of knowledge, skills and equipment as you've learned about microlaunchers over the years. You could offer to operate missions for other people.

The Microlaunchers Vision Restated

In summary, the Microlaunchers idea is that private individuals working singly and together can phenomenally change space exploration and development by starting small and building upon capabilities over time. We've reviewed the theory and technology necessary to make that revolution occur. We've reviewed ideas of how you can get involved and a vision for how things will change. This book is a call to all who want create this new future. It will start with you. To change how space is done, for access to space, first build rockets.

Appendix A - Combustion Data

The following appendix contains graphs of combustion characteristics for three different propellants at different combustion pressures and different mixture ratios.

These graphs are intended to provide a basis for experimenting with the equations in this book should you lack access to a combustion analysis software. It also provides a broad overview of combustion characteristics across mixture ratios.

There are a few trends which become obvious across the different propellants and mixture ratios. First, higher chamber pressure results in higher Isp. Second, there is a peak mixture ratio at which the propellants provide maximum Isp. What is also worth observing is that Isp generally peaks before the peak temperature and in the fuel-rich side of mixture ratios. Finally, the peak Isp mixture ratio shifts with pressure, generally, towards a leaner mixture at higher pressure.

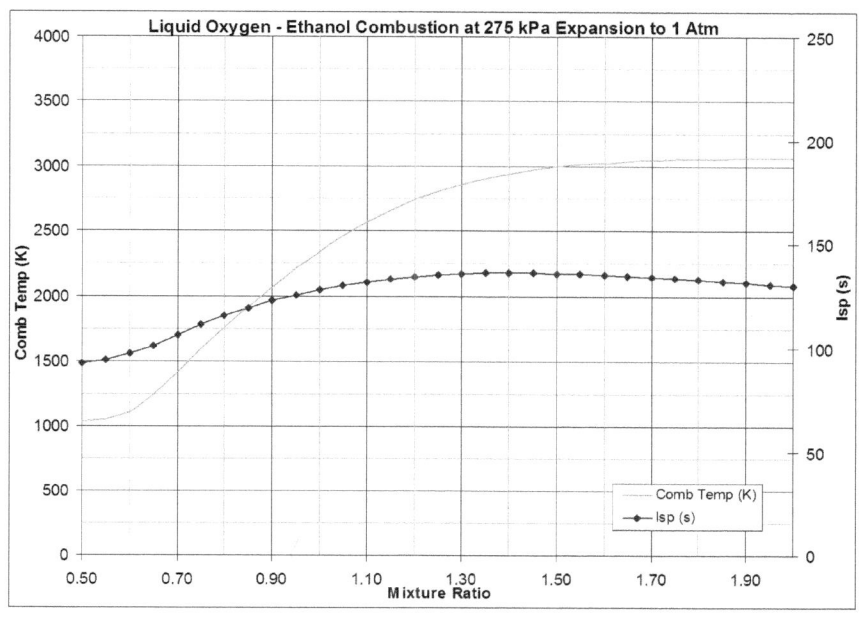

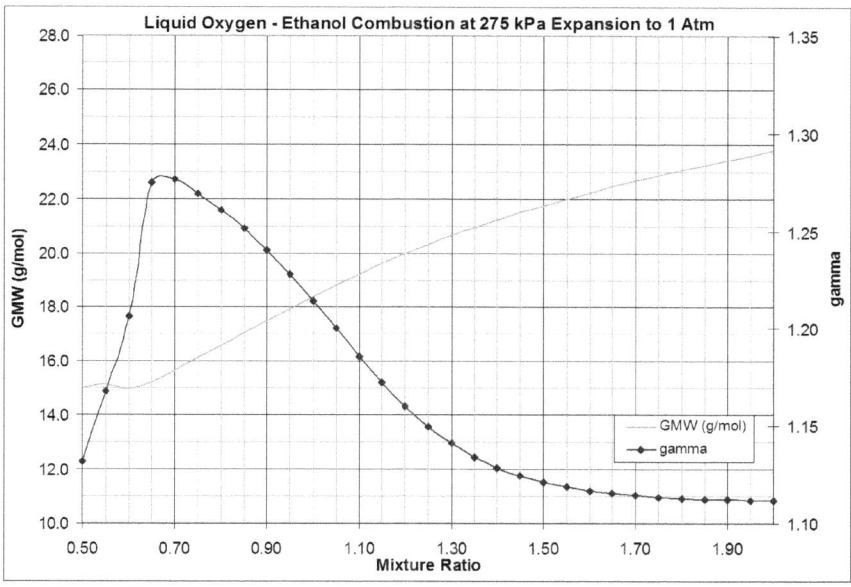

175

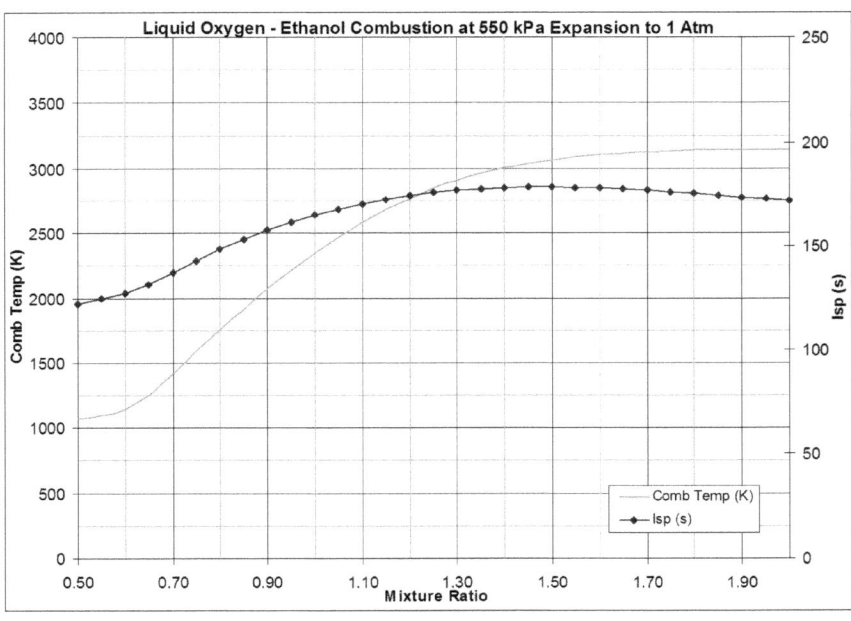

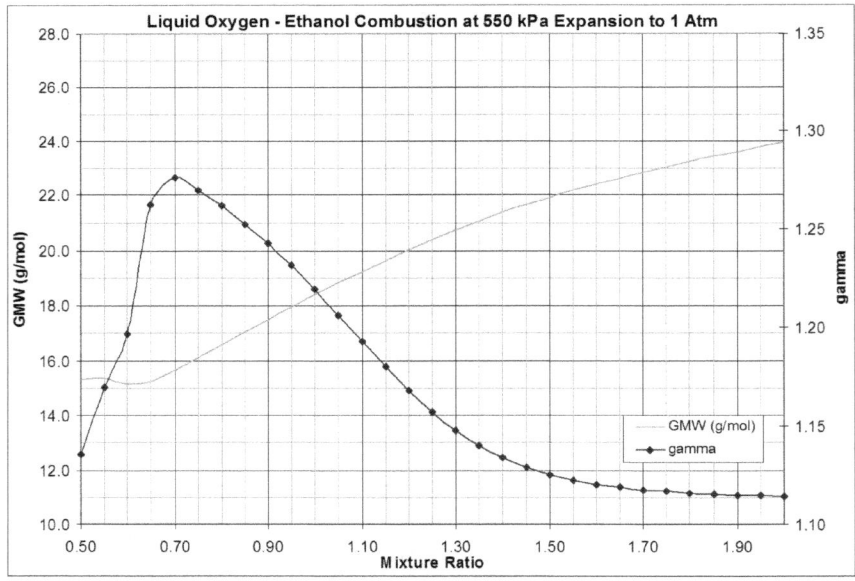

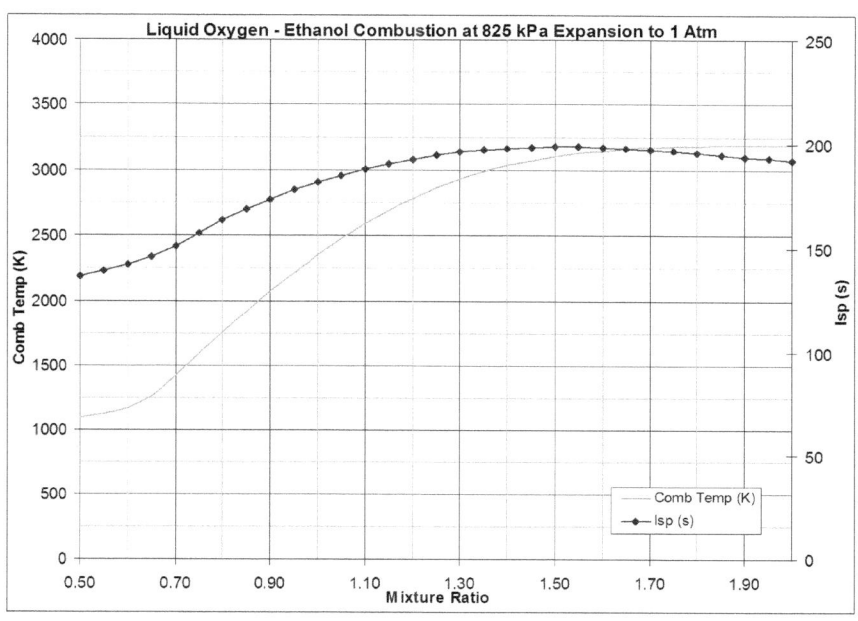

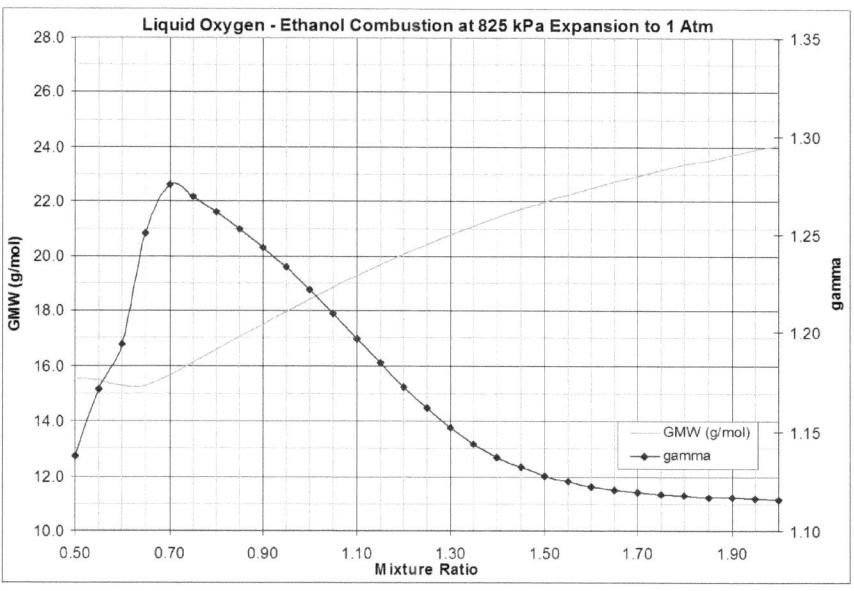

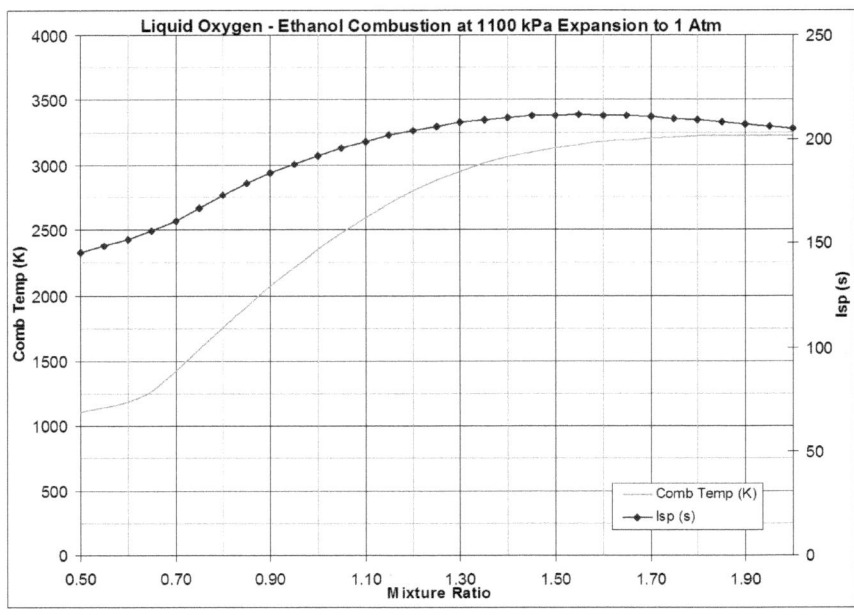

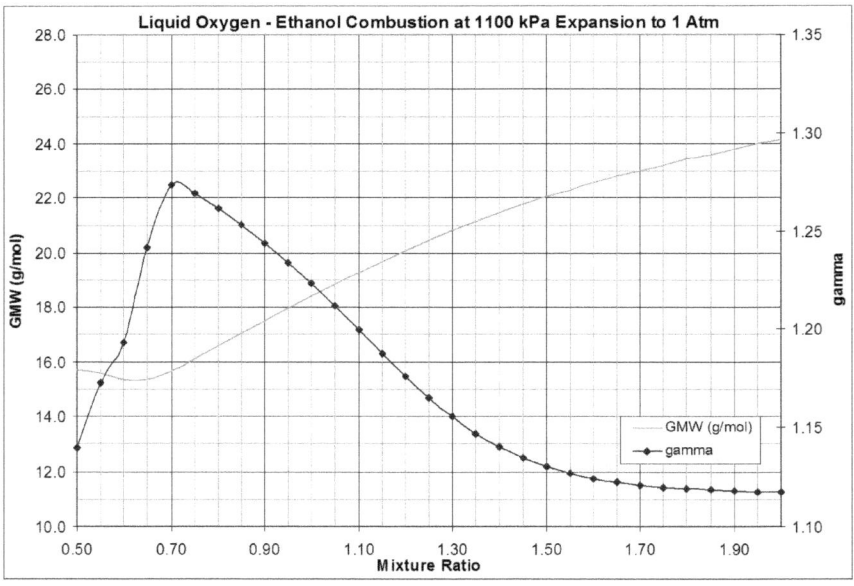

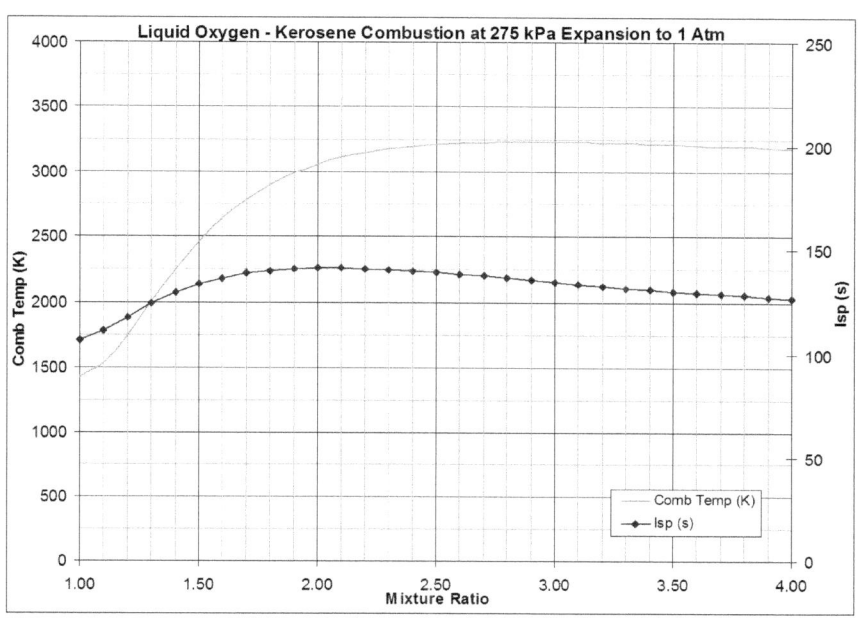

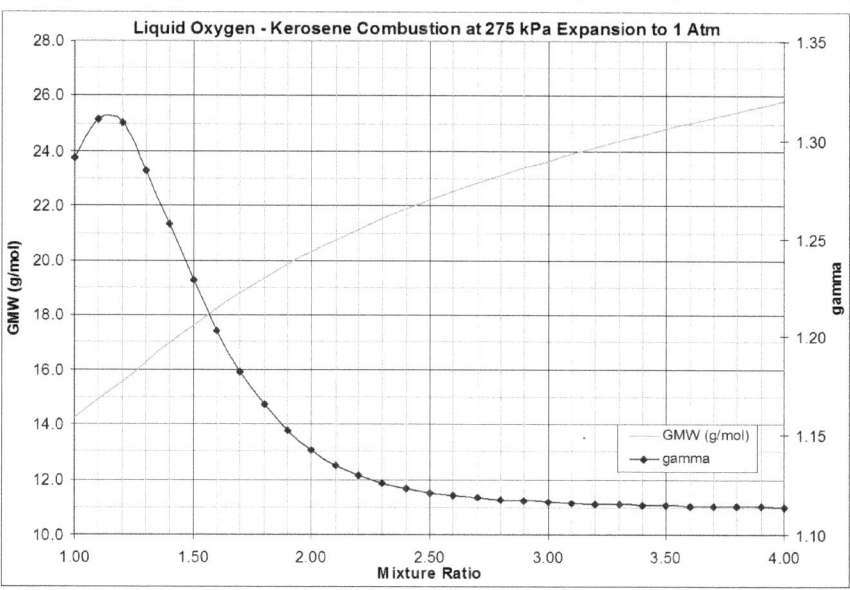

179

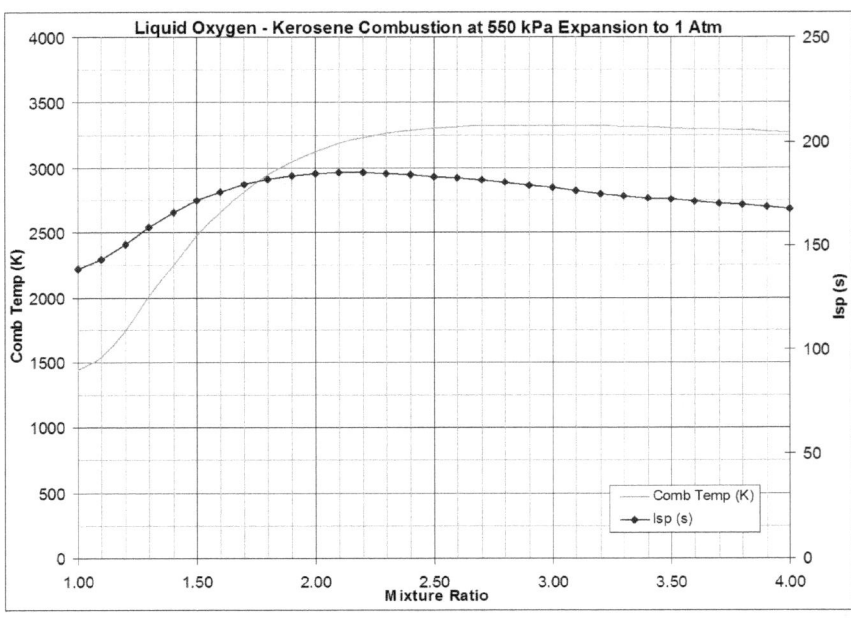

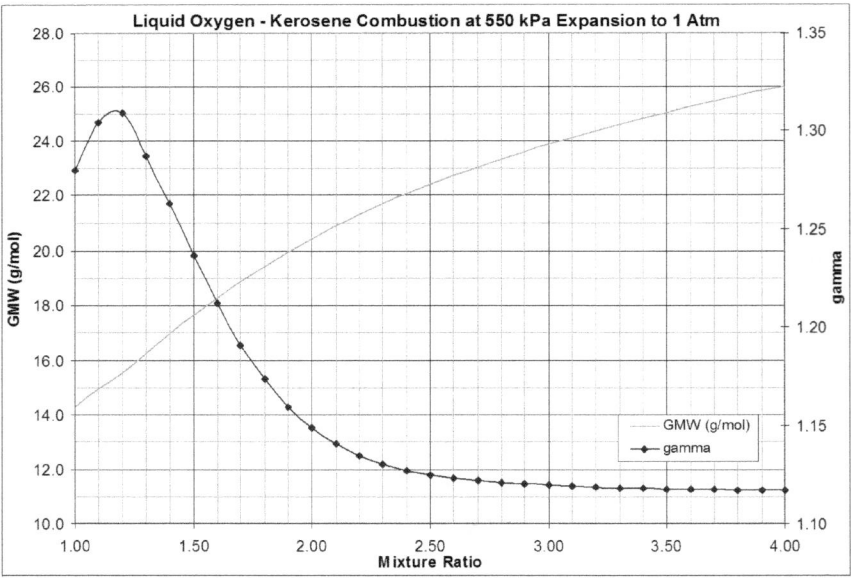

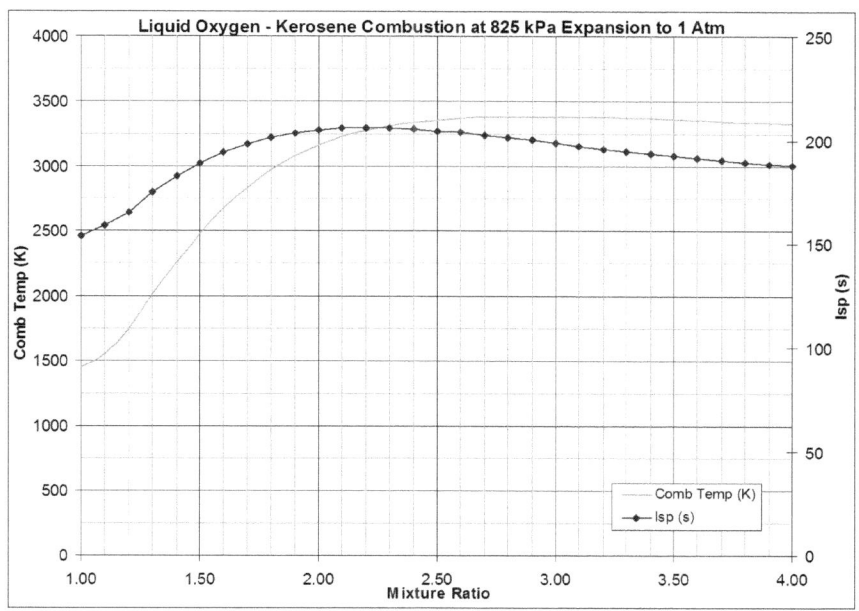

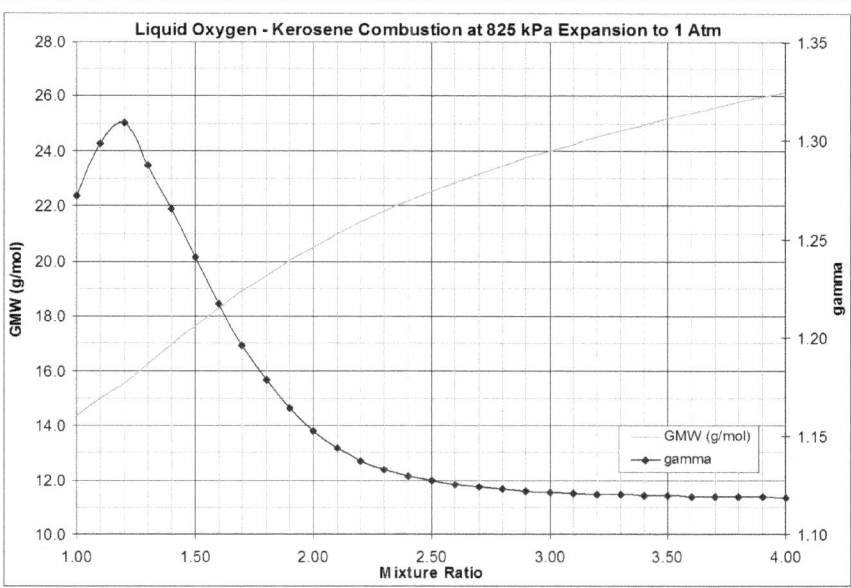

181

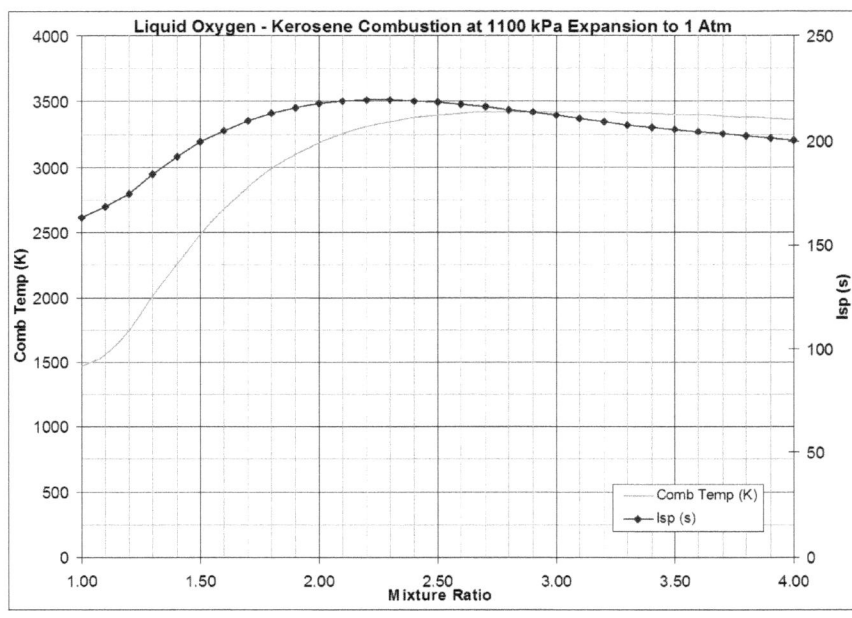

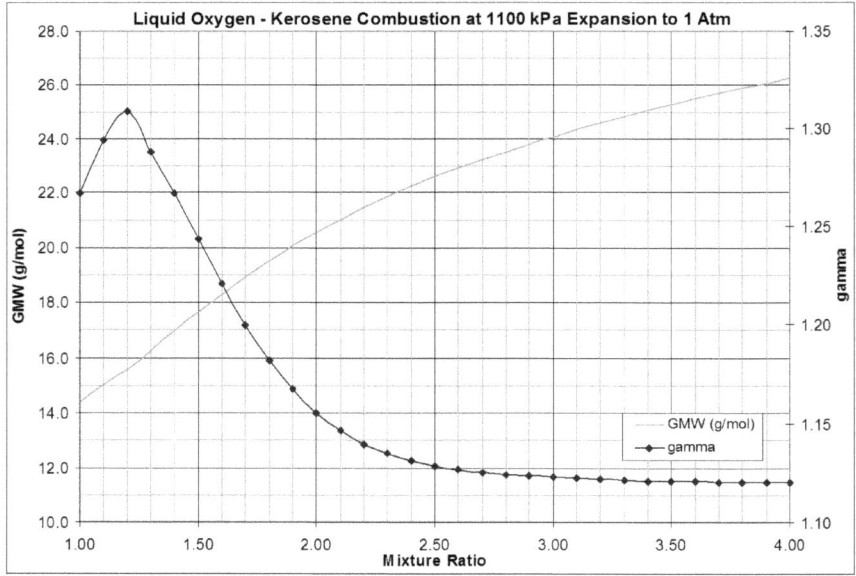

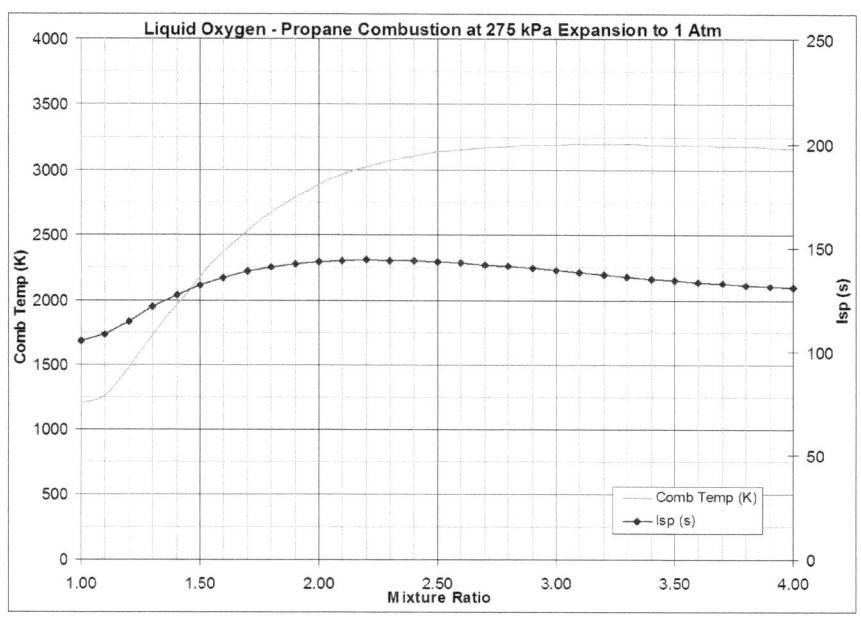

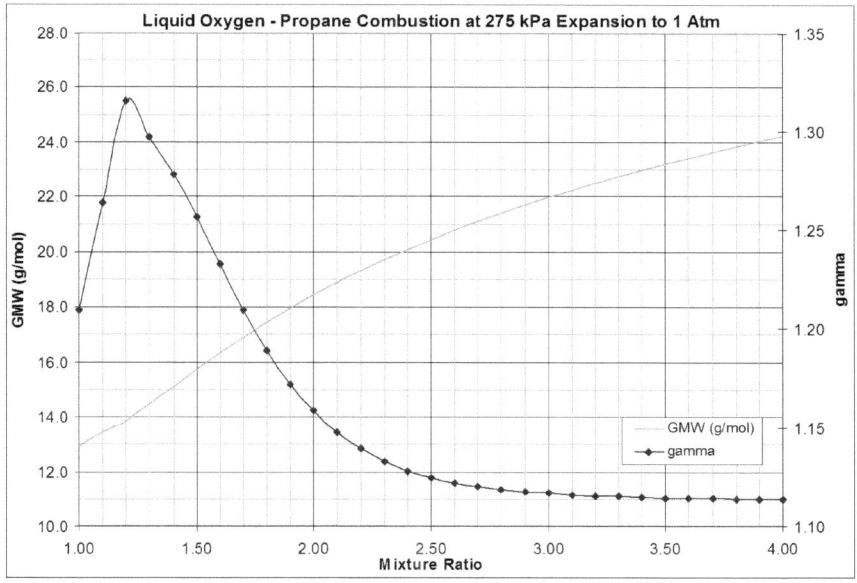

183

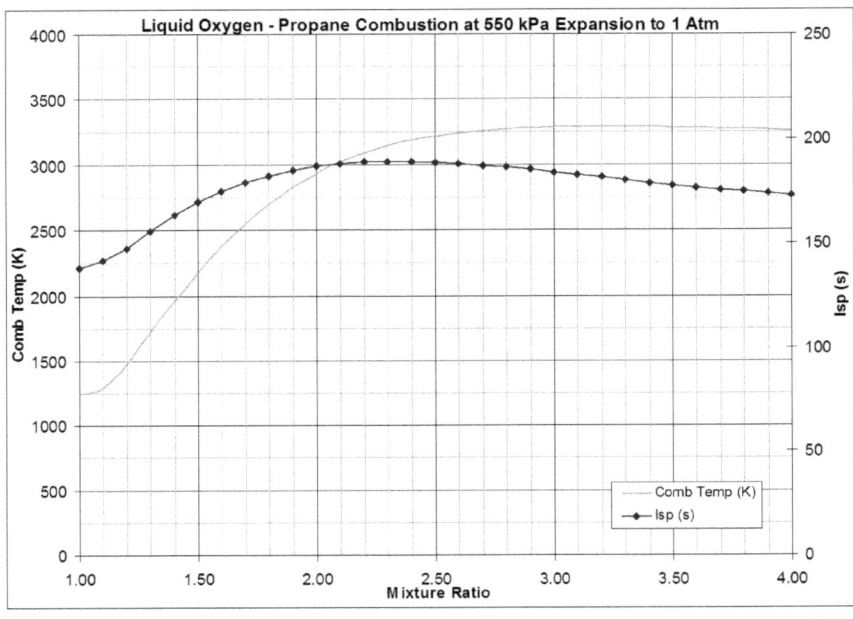

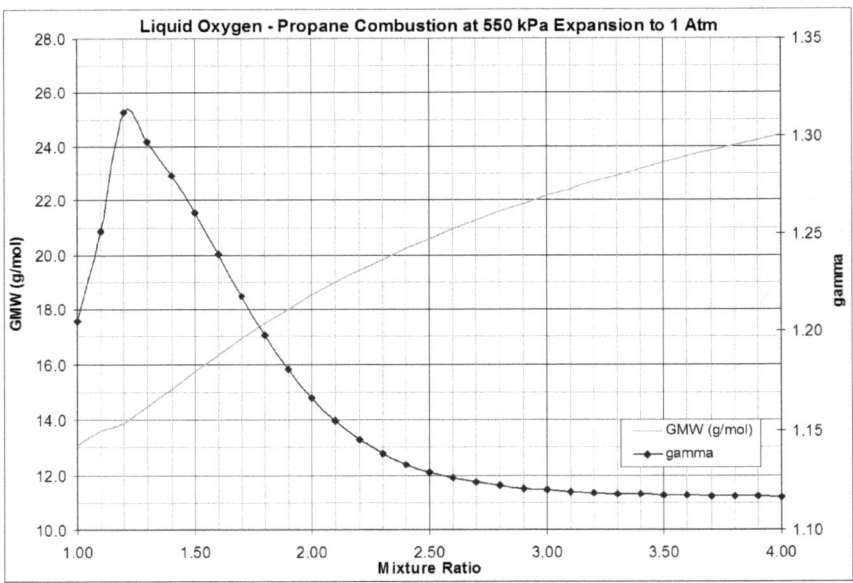

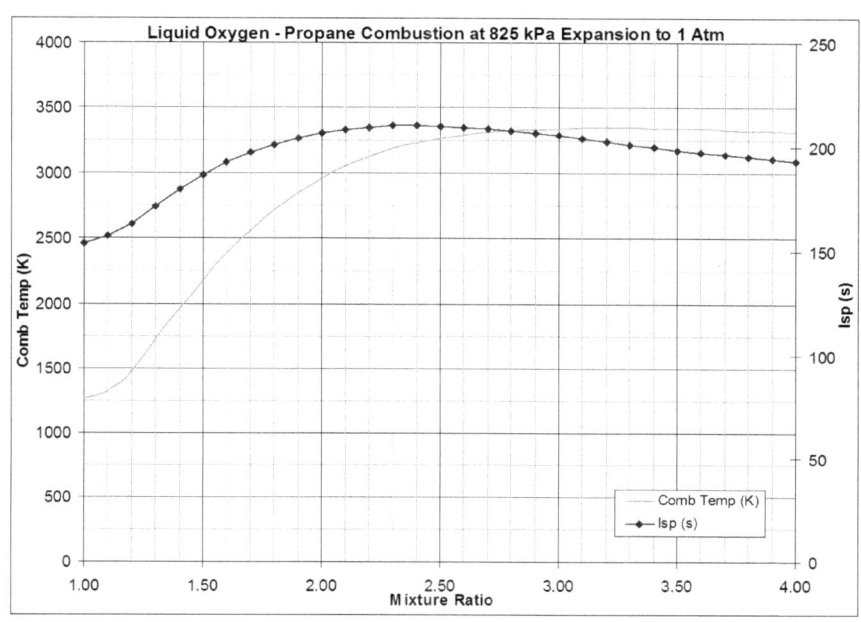

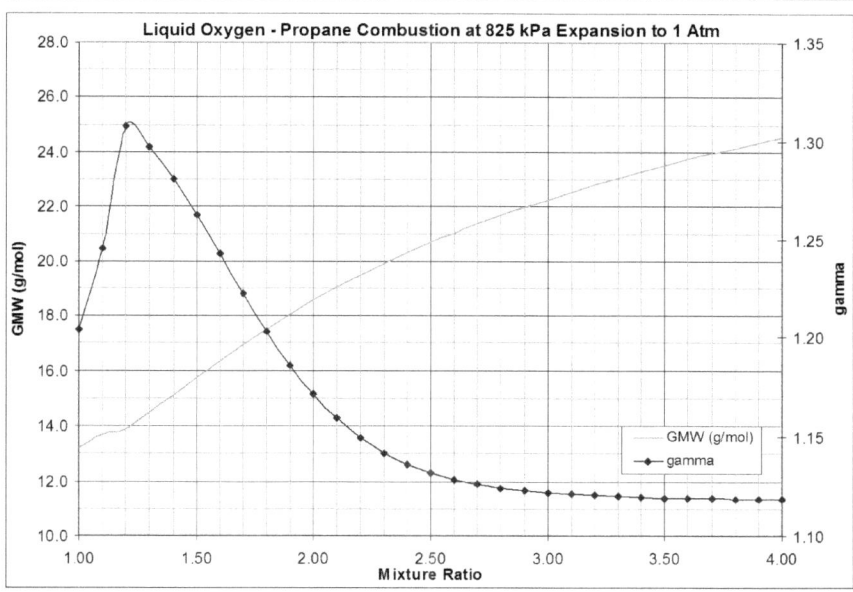

185

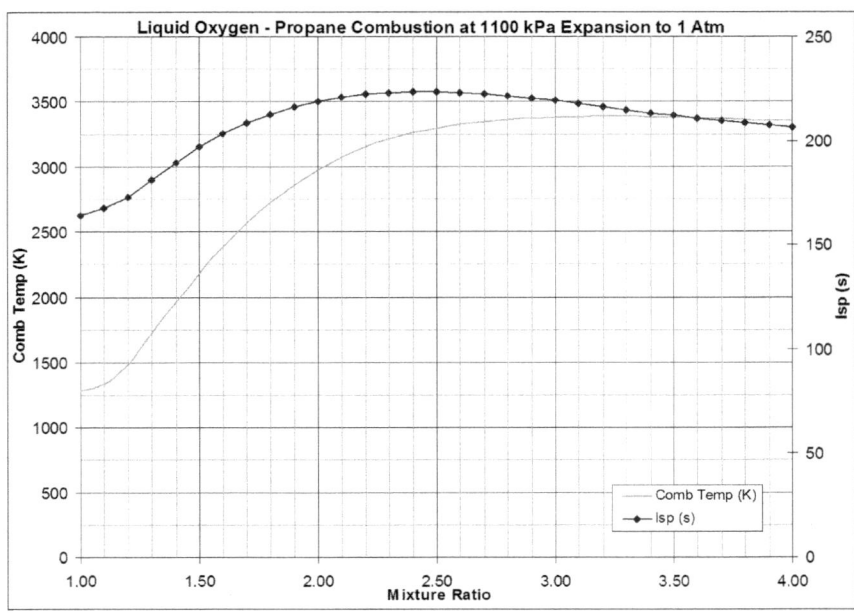

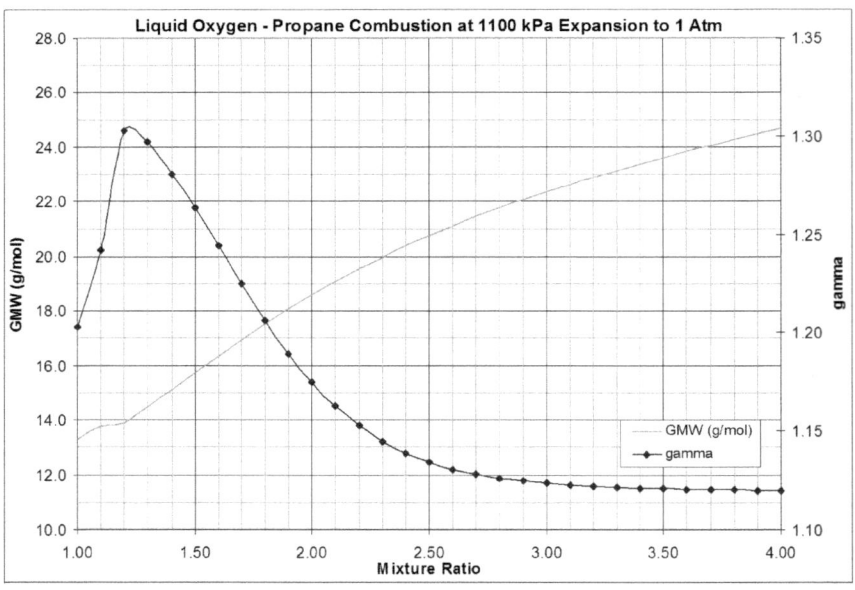

Appendix B - A Standard Atmosphere

Altitude (km)	Temperature (°C)	Pressure (kPa)	Density (kg/m³)	Speed of Sound (m/s)
0	15.000	101.325	1.22500	340.294
2.5	-1.244	74.691	0.95695	330.560
5.0	-17.474	54.048	0.73643	320.529
7.5	-33.693	38.299	0.55719	310.175
10.0	-49.898	26.499	0.41351	299.463
12.5	-56.500	16.579	0.26660	295.070
15.0	-56.500	12.111	0.19476	295.070
17.5	-56.500	8.1822	0.087340	295.070
20.0	-56.500	5.5293	0.088910	295.070
22.5	-54.079	3.7455	0.059563	296.767
25.0	-51.598	2.5492	0.040084	298.455
27.5	-49.118	1.17429	0.027103	330.133
30.0	-46.641	1.1970	0.018410	301.803
32.6	-43.286	0.81369	0.013555	304.058
35.0	-36.637	0.57459	0.0084634	308.649
37.6	-29.439	0.39857	0.0056974	312.96
40.0	-22.800	0.28714	0.0039957	317.19
42.6	-15.615	0.20329	0.0027500	321.71
45.0	-8.986	0.14910	0.0019663	325.82
47.6	-2.500	0.10751	0.0013839	329.80
50.0	-2.500	0.079779	0.0010269	329.80
52.5	-5.496	0.058438	0.00076061	328.81
55.0	-12.379	0.042525	0.00056810	322.72
57.5	-19.257	0.030691	0.0042112	319.43
60.0	-26.129	0.021958	0.0030968	315.07
62.5	-32.996	0.015567	0.0022582	310.66
65.0	-39.858	0.010929	0.00016321	306.19
67.5	-46.714	0.0075953	0.00011685	301.66
70.0	-53.565	0.0039168	0.000082829	297.06
72.5	-59.865	0.0026612	0.000057951	292.77
75.0	-64.751	0.0023881	0.000039921	289.40
77.5	-69.633	0.0011948	0.000027267	285.99
80.0	-74.511	0.00078942	0.000018458	282.54
82.5	-79.386	0.00051640	0.000012378	279.05
85.0	-84.257	0.00044568	0.0000082196	275.52
87.5	-86.28	0.00028602	0.000005328	-
90.0	-86.28	0.00018359	0.000003416	-
92.5	-86.07	0.00011798	0.000002188	-
95.0	-84.73	0.000075966	0.000001393	-
97.5	-82.11	0.000049122	0.0000008842	-
100.0	-78.07	0.000032011	0.0000005604	-

Index

A

Microlaunchers

www.ingramcontent.com/pod-product-compliance
Lightning Source LLC
Chambersburg PA
CBHW051459170526
45166CB00001B/304